Fit for hot and cold rolling of strips

Basics

Fit for hot and cold rolling of strips

Basics

by

Michael Degner and Heinz Palkowski

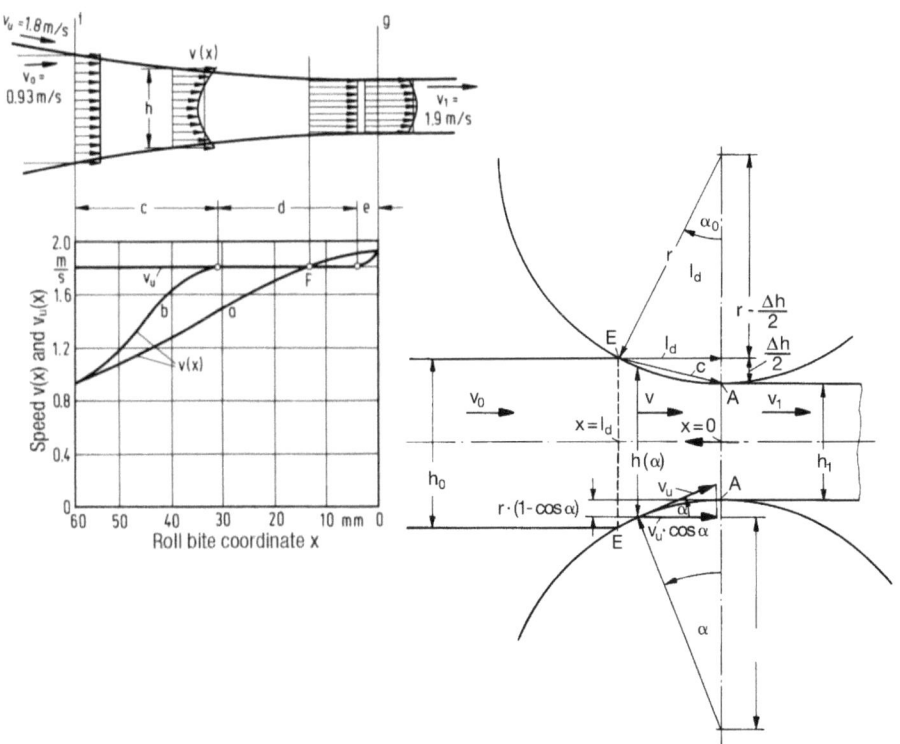

Figure on cover page according to /1, 9/

Bibliografische Information der Deutschen Nationalbibliothek:
Die Deutsche Nationalbibliothek verzeichnet diese Publikation in der Deutschen Nationalbibliografie; detaillierte bibliografische Daten sind im Internet über http://dnb.dnb.de abrufbar.

© 2016 Michael Degner and Heinz Palkowski

Illustration: Paul Westerwalbesloh

Herstellung und Verlag: IMP InterMediaPartners GmbH, Wuppertal

ISBN: 978-3-9817904-0-5

– Contents –

Preface and introduction 7

0. Abbreviations 9

1. Basic nomenclature in plasto mechanics 11
   1.1 Stress 11
   1.2 Deformation and the law of volume constancy 15
   1.3 Deformation rate... 18
   1.4 Conditions for the equilibrium 18
   1.5 Yield criterions.. 20
   1.6 Stress-deformation-relationship 23

2. Geometry of the roll gap 25

3. Biting and rolling condition 33

4. Kinematics of the roll gap.. 39

5. Application of the volume constancy law in flat rolling
   process... 50
   5.1 Calculation of production in strip rolling................... 52
   5.2 Calculation of production in reversing rolling........... 57

6. Spreading 58

7. Roll flattening and minimum strip thickness during
   cold rolling 62

8. Flow curves of metallic materials 67
   8.1 Parameters influencing the yield stress.................... 68
   8.2 Mathematical modeling of flow curves...................... 74

9. Application of flow curves and deformation energy 81

10. Elementary rolling equation 85
    10.1 Elementary rolling equation according to Siebel .. 88
    10.2 Elementary rolling equation according to von Kármán 93
    10.3 Spezific pressure distribution in the roll gap 96

11. Rolling force, torque and power – general considerations 100

12. Calculation of rolling force and rolling torque according to Lippmann and Mahrenholtz 105

13. Calculation of rolling force and torque for hot rolling 108

14. Calculation of rolling force and torque for cold rolling 114

15. Examples for the calculation of rolling forces and torques in hot and cold rolling 120

16. Heat balance in hot rolling and temperature calculation 123

17. Literature 131

18. List of key words 134

Preface and introduction

This book is written for students at technical universities and engineers working in rolling mills. It can be used for their individual examination preparations and for upgrading basic knowledge in rolling theory. Latter is necessary for solving daily challenges in practical work. Some equations are compiled to provide practical estimations of rolling parameters to be used in daily work.

Crude steel production in the world has been raised to roughly 1.7 bn tons in 2014. In hot and cold flat rolling mills a part of roughly 70 % is further processed to semi- and finished products used e.g. in automotive, house packaging, ship building and construction industry in the developed countries in the world. Average production of further processed flat products in the world is roughly 50 % of the complete crude steel production.

Flat rolling is a forming method aiming in reduction the cross-section area of a work piece (e.g. semi-finished ingots), enlarging its length and adjusting its properties like geometry, ductility, strength and surface finish structure.

In the flat rolling process the rolls are cylinder-like and the cross-section of the rolled material is rectangular.

The material is squeezed from height in prior length direction, but partly as well in width direction. The width flow direction is undesired in many applications. Under the assumption that width size is larger than tenfold height size the rolling pocess can be considered to be without material spreading (plain deformation). In this case the stripe model of the elementary plasticity theory can be applied for calculation of the integral parameters like roll force, roll torque and roll power.

Regarding material insert temperature it will be distinguished between hot and cold rolling. Hot rolling is performed above the recrystallization temperature of the material. Cold rolling usually takes place at ambient temperature. But the material temperature can increase due to dissipation effects during cold rolling. In the case of flat rolling the material is hot rolled in the first rolling steps. This is because of the reduced roll forces needed and the increased material formability at high temperature levels. Disadvantages of hot rolling are the high energy use and scale formation. Due to heat loss of the material hot rolling is not practicable for thin strip. In this case cold rolling is performed. Cold rolling in the process chain of an integrated steel works improves the material surface finish and material properties.

This book gives the basic equations and nomenclature of hot and cold rolling technology including the derivation from physical laws like mass and energy constancy and the third Newton axiom. These equations are used for the exercises given in "Fit for hot and cold rolling of strips".

0. Abbreviations

In this book the following abbreviations are used:

a	[-]	lever arm
a_{su}	[m s^{-2}]	acceleration, speed up
A	[mm^2]	area
b	[mm]	width
c_p	[J kg^{-1} K^{-1}]	specific heat capacity
C_H	[mm^2 N^{-1}]	Hitchcock's work roll flattening constant
d	[mm]	roll diameter
F	[N]	force
h	[mm]	thickness
k, k_f, k_{fm}	[N mm^{-2}]	yield stress
l	[mm, m]	length
l_d	[mm]	roll bite length
m	[kg, t]	mass
n	[min^{-1}]	roll revolution
M	[N m]	roll torque
P	[kg s^{-1}, t h^{-1}]	productivity
s_H	[N mm^{-2}]	average normal stress
T	[K, °C]	temperature
r	[mm]	roll radius
t	[s, min, h, a]	time
v	[m s^{-1}, m min^{-1}]	speed
V	[mm^3]	volume
x_n	[mm]	neutral point
α	[°, rad]	angle coordinate
α_N	[°, rad]	neutral angle
β	[-]	spreading ratio
ε	[%]	relative deformation, relative reduction

δ	[-]	factor
Δb	[mm]	width reduction
Δh	[mm]	height reduction, thickness reduction
κ	[-], [%]	slip forward
σ	[N mm^{-2}]	normal stress
σ_B	[N mm^{-2}]	backward tension
σ_F	[N mm^{-2}]	forward tension
σ_N	[N mm^{-2}]	normal stress
$\sigma_1, \sigma_2, \sigma_3$	[N mm^{-2}]	main normal stress
τ	[N mm^{-2}]	shear stress
ρ	[t m^{-3}]	density
μ	[-]	friction coefficient
φ	[-]	logarithmic deformation, true strain
$\dot{\varphi}$	[s^{-1}]	deformation rate
γ	[-]	compression ratio
γ_R	[rad]	friction angle
λ	[-]	length ratio
$\lambda(T)$	[W K^{-1} mm^{-1}]	specific heat capacity
ϖ	[J mm^{-3}]	energy density

Indices
0   entry/start
1   exit/end
e   exit
m   average
u   peripheral
x, y, z   coordinates x, y, z

1. Basic nomenclature in plasto mechanics

1.1 Stress

Stress is defined as the acting force per area unit. The simplest case is the one axis strain characterized by an acting force rectangular to the work piece surface area. In practice this is not the case. Generally the outer deformation forces act not only in one direction but in all three directions. Forces and areas have to be considered mathematically as vectors. In case of spatial stress conditions the situation is not really simple. Each vector is described by three components. A force ΔF acting on an area element ΔA results in general in nine values for the quotient force/area. This is one element of a (3x3)-matrix with the mathematical form in a Cartesian system:

$$\begin{pmatrix} \dfrac{\Delta F_x}{\Delta A_x} & \dfrac{\Delta F_y}{\Delta A_x} & \dfrac{\Delta F_z}{\Delta A_x} \\ \dfrac{\Delta F_x}{\Delta A_y} & \dfrac{\Delta F_y}{\Delta A_y} & \dfrac{\Delta F_z}{\Delta A_y} \\ \dfrac{\Delta F_x}{\Delta A_z} & \dfrac{\Delta F_y}{\Delta A_z} & \dfrac{\Delta F_z}{\Delta A_z} \end{pmatrix}. \quad (1.1)$$

Taking into account the limit conditions for the areas nine stress values follow:

$$\lim_{\Delta A_x \to 0} \frac{\Delta F_x}{\Delta A_x} = \sigma_x$$

$$\lim_{\Delta A_x \to 0} \frac{\Delta F_y}{\Delta A_x} = \tau_{xy} \qquad (1.2)$$

$$\lim_{\Delta A_x \to 0} \frac{\Delta F_z}{\Delta A_x} = \tau_{xz}$$

i.e. three normal and three shear stresses, **Fig. 1.1**:

$$\begin{pmatrix} \sigma_x & \tau_{xy} & \tau_{xz} \\ \tau_{yx} & \sigma_y & \tau_{yz} \\ \tau_{zx} & \tau_{zy} & \sigma_z \end{pmatrix} \qquad (1.3)$$

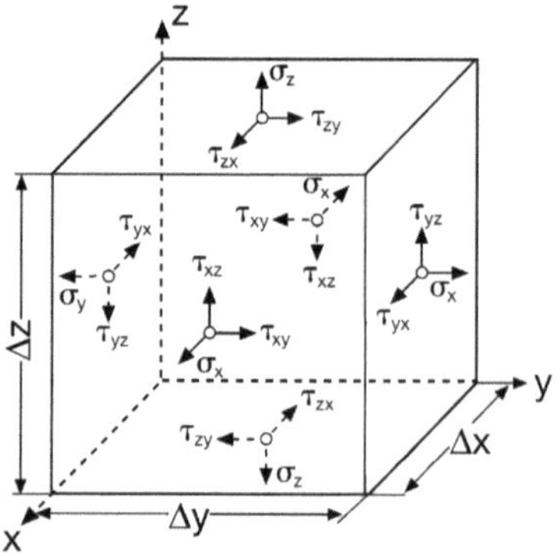

Fig. 1.1: Stresses at a volume element according to /9/

In case of a normal stress σ the force acts rectangular to the impacted area, i.e. force and normal area direction are identical. The shear stress is characterized by the direction of the normal area (1st index) rectangular to the force direction (2nd index). I.e. the force acting line is within the impacted area. These resulting nine stresses can be written as a matrix which is named stress tensor:

$$\sigma_{ij} = \begin{pmatrix} \sigma_x & \tau_{yx} & \tau_{zx} \\ \tau_{xy} & \sigma_y & \tau_{zy} \\ \tau_{xz} & \tau_{yz} & \sigma_z \end{pmatrix}. \qquad (1.4)$$

Assuming that the sum of the inner moments equalizes to zero, the moment equilibrium at the volume element is e.g. for the z-axis

$$\tau_{xy} \cdot \Delta y \cdot \Delta z \cdot \Delta x = \tau_{yx} \cdot \Delta x \cdot \Delta z \cdot \Delta y \qquad (1.5)$$

and therefore $\tau_{xy} = \tau_{yx}$. In the same way all other shear stresses can be treated, so all indices can be changed in the following way:

$$\tau_{xy} = \tau_{xy}, \; \tau_{yz} = \tau_{zy}, \; \tau_{zx} = \tau_{xz}. \qquad (1.6)$$

Generally an acting stress is characterized by three normal and three shear stresses.

In case the section plane in a body is chosen in the way that all shear stresses vanish, these stress planes are called **main stress areas**. This is a mathematical treatment and can be attained by rotating the coordinate system in the way that only the normal stresses act, the so called **main normal stresses** $\sigma_1, \sigma_2, \sigma_3$. By convention it is:

$$\sigma_1 \geq \sigma_2 \geq \sigma_3.$$

The maximum shear stresses are angled at 45° to the direction of the main stress. Relevant for the yield criterion is the absolute maximum shear stress:

$$\tau_{max} = \frac{1}{2} \cdot (\sigma_1 - \sigma_3). \tag{1.7}$$

A stress with identical normal stress in all directions and no shear stress is named **hydrostatic stress**. This stress provokes only elastic deformation on the body, so no plastic deformation. The hydrostatic part s_H of a three-dimensional stress situation can be written under the assumption that the same elastic volume change is provoked as under the given stress:

$$s_H = \frac{1}{3} \cdot (\sigma_x + \sigma_y + \sigma_z). \tag{1.8}$$

This is the average of the three normal stress values. Therefore s_H is denoted as **average normal stress**. The hydrostatic stress

$$\begin{pmatrix} s_H & 0 & 0 \\ 0 & s_H & 0 \\ 0 & 0 & s_H \end{pmatrix} \tag{1.9}$$

is not influencing the plastic deformation. Therefore it will be subtracted from the stress tensor if calculating the flow condition. The remaining part is named **stress deviator** and is described by

$$\begin{aligned} s_{Hx} &= \sigma_x - s_H, \\ s_{Hy} &= \sigma_y - s_H, \\ s_{Hz} &= \sigma_z - s_H. \end{aligned} \tag{1.10}$$

The sum of the normal components vanishes:

$$S_{Hx} + S_{Hy} + S_{Hz} = 0. \tag{1.11}$$

1.2 Deformation and the law of volume constancy

Exterior forces acting on a body provoke interior stresses within the body. They result in elastic and plastic deformations. The elastic deformation disappears when the forces are removed. In case of none vanishing reinforced deformation a plastic deformation takes place. The total deformation acting on the body is the sum of elastic and plastic deformation:

$$\varepsilon = \varepsilon_e + \varepsilon_{pl}. \tag{1.12}$$

If a body is compressed on one axis from its initial height h_0 to its final height h_1 the total height reduction can be written as:

$$\Delta h = h_0 - h_1. \tag{1.13}$$

For the calculation of deformation the height reduction dh can be related to the initial height:

$$d\varepsilon = \frac{dh}{h_0}. \tag{1.14}$$

The related deformation totally is:

$$\varepsilon = \int_{h_1}^{h_0} \frac{dh}{h_0}$$
$$\varepsilon = \frac{h_0 - h_1}{h_0}$$
(1.15)

Another way is to relate the height reduction dh on the respective intermediate height.

In this case the related height reduction can be written as:

$$d\varphi = \frac{dh}{h}.$$
(1.16)

By integration the logarithmic deformation follows according to:

$$\varphi = \int_{h_0}^{h_1} \frac{dh}{h}$$
$$\varphi = \ln \frac{h_0}{h_1}$$
(1.17)

The total deformation applying i-deformation steps constitutes:

$$\varphi_g = \varphi_1 + \varphi_2 + \ldots + \varphi_i$$
(1.18)

or

$$\varphi_g = \ln\frac{h_0}{h_1} + \ln\frac{h_1}{h_2} + \ldots + \ln\frac{h_{i-1}}{i} = \ln\frac{h_0}{h_i}.$$
(1.19)

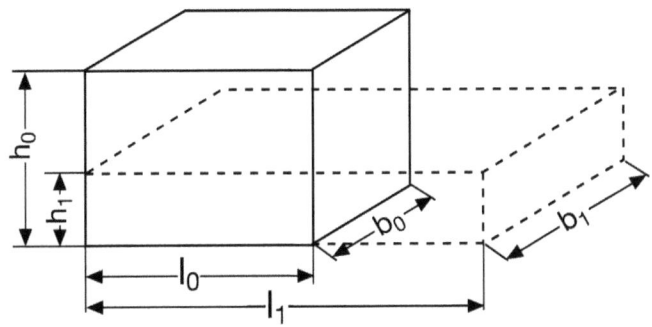

Fig. 1.2: Parallelepipedical deformation of a body according to /9/

During plastic deformation the volume of the body is assumed to be constant. The law of volume constancy while deforming a parallelepiped, **Fig. 1.2**, is:

$$h_0 \cdot b_0 \cdot l_0 = h_1 \cdot b_1 \cdot l_1 \qquad (1.20)$$

or

$$\frac{h_0 \cdot b_0 \cdot l_0}{h_1 \cdot b_1 \cdot l_1} = 1. \qquad (1.21)$$

Therefore the sum of the logarithmic deformations in height, length and width direction is always zero:

$$\varphi_h + \varphi_b + \varphi_l = 0. \qquad (1.23)$$

By knowing compression and spreading during deformation the elongation φ_l of the body can be calculated:

$$\varphi_l = -(\varphi_h + \varphi_b). \qquad (1.24)$$

When rolling a strip material spreading can be neglected if the ratio width b to thickness h is > 10, so $\varphi_b = 0$. Therefore in a first approach $\varphi_l = -\varphi_h$ is valid during strip rolling.

1.3 Deformation rate

The deformation rate is defined as the change of logarithmic deformation $d\varphi$ in a time interval dt:

$$\dot{\varphi} = \frac{d\varphi}{dt}. \qquad (1.25)$$

The deformation rates in width-, height- and length direction can be expressed according to:

$$\dot{\varphi}_b = \frac{1}{b} \cdot \frac{db}{dt}, \quad \dot{\varphi}_h = \frac{1}{h} \cdot \frac{dh}{dt}, \quad \dot{\varphi}_l = \frac{1}{l} \cdot \frac{dl}{dt}. \qquad (1.26)$$

Applying on a compression test the equation $\dot{\varphi}_h = \frac{d\varphi}{dh} = \frac{1}{h} \cdot \frac{dh}{dt} = \frac{v}{h}$ with v as compression speed is valid.

Regarding the law of volume constancy it can be deducted:

$$\dot{\varphi}_h + \dot{\varphi}_b + \dot{\varphi}_l = 0. \qquad (1.27)$$

1.4 Conditions for the equilibrium

The equilibrium condition of the classic mechanics states that the sum of forces and torques in a closed system is always zero. This physical law is not only valid for a body but as well for a volume element. Due to the equilibrium of torques the permutability of the shear stress indices result. In general the gravity and mass inertia forces are neglected in the calculations and questions of mechanical deformation. The sum of all forces result in the following equilibrium condition for the x-direction if the forces are divided by the volume element $\Delta V = \Delta x \cdot \Delta y \cdot \Delta z$:

$$\frac{\Delta \sigma_x}{\Delta x} + \frac{\Delta \tau_{yx}}{\Delta y} + \frac{\Delta \tau_{zx}}{\Delta z} = 0. \qquad (1.28)$$

The equilibrium conditions for any position within the volume element are deducted in the same manner if the acting forces in y- and z-direction are added and the limit values for the parameters $\Delta x \to 0$, $\Delta y \to 0$ and $\Delta z \to 0$ are calculated. The result of this mathematical operation is:

$$\frac{\partial \sigma_x}{\partial x} + \frac{\partial \tau_{yx}}{\partial y} + \frac{\partial \tau_{zx}}{\partial z} = 0,$$

$$\frac{\partial \tau_{xy}}{\partial x} + \frac{\partial \sigma_y}{\partial y} + \frac{\partial \tau_{zy}}{\partial z} = 0, \qquad (1.29)$$

$$\frac{\partial \tau_{xz}}{\partial x} + \frac{\partial \tau_{yz}}{\partial y} + \frac{\partial \sigma_z}{\partial z} = 0.$$

These relations are not suitable for practical applications due to their mathematical effort for solution. In the elementary rolling theory the equilibrium for a finite volume element (stripe) is considered. Important for solution of practice relevant questionnaires is the knowledge of the yield criterion.

1.5 Yield criterions

The yield criterion is a mathematical description for the beginning of the plastic deformation under uniaxial stress conditions. In the elastic forming range, which has to be overcome before plastic deformation takes place, the applied stress is proportional to the deformation according to the Hooke's law. As proportionality factor the elastic modulus (Young's modulus) E is introduced. Once the stress in its total amount overcomes the material yield strength YS, plastic deformation starts. Indeed, the total stress state plays a role because in case a cylindrical round specimen is enforced from all sides by pressure it starts flowing at a minor stress niveau than in case of exclusively applied tension in length direction. In the yield criterion beside the stress in length direction the radial pressure has to be considered, too.

According to von Mises /18/ the yield criterion is valid as following:

$$S_{Hx}^2 + S_{Hy}^2 + S_{Hz}^2 + 2 \cdot \left(\tau_{xy}^2 + \tau_{yz}^2 + \tau_{zx}^2\right) = 2 \cdot k^2 = \frac{2}{3} \cdot k_f^2, \quad (1.30)$$

where k_f denotes the yield stress of an ideal-plastic body if only pure shear stresses act. Considering an istropic material the yield criterion is invariant, i.e. there is no dependency from the chosen position of the coordinate system. For the main stress directions the yield criterion can be written as:

$$S_{Hx}^2 + S_{Hy}^2 + S_{Hz}^2 = 2 \cdot k^2. \quad (1.31)$$

Reducing the deviator stress to the stress itself provides:

$$\left(\sigma_1^2 - \sigma_2^2\right)^2 + \left(\sigma_2^2 - \sigma_3^2\right)^2 + \left(\sigma_3^2 - \sigma_1^2\right)^2 = 6 \cdot k^2. \qquad (1.32)$$

Applying tension or pressure load on one axis with $\sigma_1 = k_f$ and $\sigma_2 = \sigma_3 = 0$ the following equation is valid:

$$\sigma_1 = \sqrt{3} \cdot k. \qquad (1.33)$$

According to the Mises theory the yield stress is $\sqrt{3}$ larger in the case of uniaxial load than in case of pure shear.

Taking into account the above equations the yield criterion can be formulated for the general three-dimensional stress case to:

$$k_f = \sqrt{\frac{1}{2} \cdot \left[\left(\sigma_1 - \sigma_2\right)^2 + \left(\sigma_2 - \sigma_3\right)^2 + \left(\sigma_3 - \sigma_1\right)^2\right]}. \qquad (1.34)$$

For an arbitrary rotated rectangular coordinate system it follows:

$$k_f = \sqrt{\frac{1}{2} \cdot \left[\left(\sigma_1 - \sigma_2\right)^2 + \left(\sigma_2 - \sigma_3\right)^2 + \left(\sigma_3 - \sigma_1\right)^2\right] + 3 \cdot \left(\tau_{xy}^2 + \tau_{yz}^2 + \tau_{zx}^2\right)}. \qquad (1.35)$$

This yield criterion indicates that the volume starts yielding if the deformation energy achieves a critical value. Therefore it is named deformation energy hypothesis.

In the elementary plastomechanics the yield criterion according to Tresca is used where the metallic stock starts yielding if the maximum main stress difference of the spatial stress condition achieves a critical value of the shear stress:

$$k = \tau_{max} = \frac{1}{2} \cdot k_f \qquad (1.36)$$

and therefore it follows

$$k_f = \sigma_1 - \sigma_3 . \qquad (1.37)$$

Assumption for the application of this yield criterion is that the coordinates of the main stresses are known.

According to Tresca the start of yielding is determined by the maximum and minimum main stresses σ_1 and σ_3; von Mises further takes the influence of the main stress σ_2 into account. The largest influence of σ_2 is given if $\sigma_2 = \frac{1}{2} \cdot (\sigma_1 + \sigma_3)$ is valid. This is the case of plane deformation. This assumption can be made in the case of spreading-free strip rolling. Inserting this arithmetic average in the Mises yield criterion gives:

$$\sigma_1 - \sigma_3 = \frac{2}{\sqrt{3}} \cdot k_f = 1{,}155 \cdot k_f . \qquad (1.38)$$

By comparison with the yield criterion according to Tresca there is a difference of maximum 15% depending on the stress condition.

1.6 Stress-deformation relationship

For complete modeling of the elastic and plastic behavior of a body the yield criterion and the relation between stress and strain must be known.

In case of elastic deformation the Hooke's law is to be applied. The deformation is proportional to the acting stress. In case of uniaxial stress condition it is:

$$\varepsilon_x = \frac{1}{E} \cdot \sigma_x .\qquad(1.39)$$

In the plastic region the stress σ_x is to be substituted by the deviator stress s_{Hx} and the elastic deformation ε_x by the deformation increment $d\varepsilon_x$. Introducing the proportionality factor $d\delta$ instead of $1/E$ gives for the deformation in x-direction:

$$d\varepsilon_x = d\delta \cdot s_{Hx} .\qquad(1.40)$$

Dividing by the time increment dt provides for an ideal plastic body (no strain- hardening) the strain speed. The strain speed is proportional to the deviator stress:

$$\dot{\varepsilon}_x = \dot{\sigma} \cdot s_{Hx} .\qquad(1.41)$$

The other deformations are to be treated analogue to obtain the stress-deformation-strain relation or the material law according to Levy or material law according to Mises:

$$\dot{\varepsilon}_x = \dot{\delta} \cdot S_{Hx}, \quad \dot{\varepsilon}_y = \dot{\delta} \cdot S_{Hy}, \quad \dot{\varepsilon}_z = \dot{\delta} \cdot S_{Hz}, \tag{1.42}$$

$$\frac{1}{2} \cdot \dot{\gamma}_{xy} = \dot{\delta} \cdot \tau_{xy}, \quad \frac{1}{2} \cdot \dot{\gamma}_{yz} = \dot{\delta} \cdot \tau_{yz}, \quad \frac{1}{2} \cdot \dot{\gamma}_{zx} = \dot{\delta} \cdot \tau_{zx}. \tag{1.43}$$

The proportionality factor $\dot{\delta}$ depends on the material itself and on all components of the strain speed, i.e. the factor $\dot{\delta}$ is not constant. Using the yield criterion according to Mises it follows:

$$\dot{\delta} = \frac{1}{k_f} \cdot \sqrt{\frac{3}{2} \cdot \left(\dot{\varepsilon}_x^2 + \dot{\varepsilon}_y^2 + \dot{\varepsilon}_z^2 \right) + \frac{3}{4} \cdot \left(\dot{\gamma}_{xy}^2 + \dot{\gamma}_{yz}^2 + \dot{\gamma}_{zx}^2 \right)}. \tag{1.44}$$

It can be stated: In the elastic region stress and deformation are proportional and in the plastic region stress and deformation rate are.

For solving the general deformation problem there are ten basic equations available, named the equilibrium conditions, the yield criterion and the material law. Unknown parameters are the six deformation stresses (three normal stresses and three shear stresses) σ_x, σ_y, σ_z, τ_{xy}, τ_{yz}, τ_{zx}, the three speeds v_x, v_y, v_z and the factor $\dot{\delta}$ of the material law. An analytical solution of this equation normally does not exist. In metal forming technology approximate solutions (elementary plasto mechanics) are often used and applied.

2. Geometry of the roll gap

Rolling of sheet strip or plate material is performed with – at first glance – cylindrical rolls. The material is reduced in thickness being elongated in length. Spreading can be neglected if the ratio width to thickness is larger than ten. **Fig. 2.1** gives an overview of the roll gap geometry.

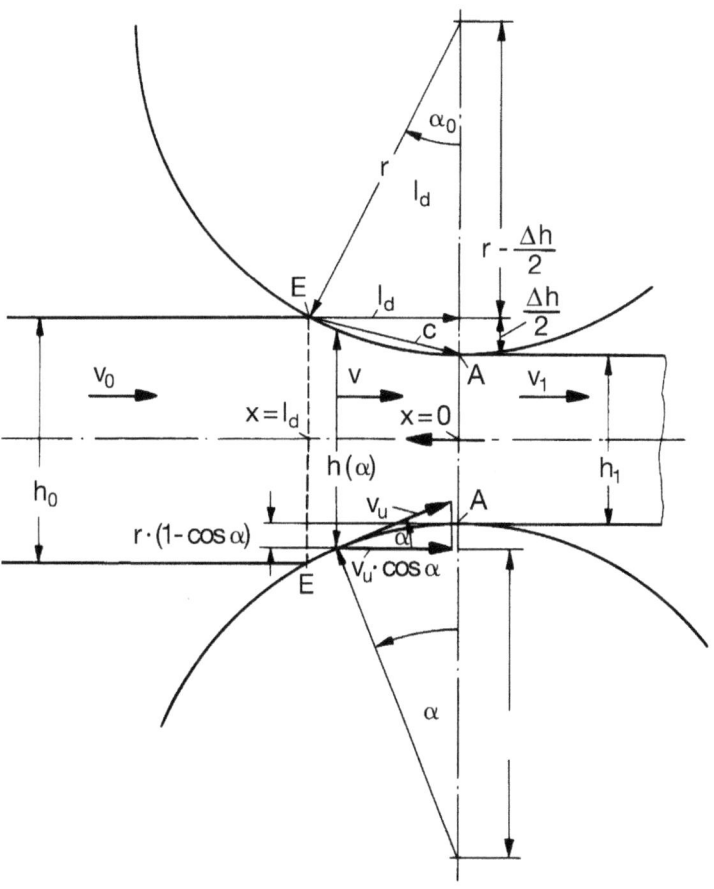

Fig. 2.1: Nomenclature of the roll gap - overview /1, 9/

For the calculation the following definitions and terms will be used:

b_0	[mm]:	Entry width.
b_1	[mm]:	Exit width.
$b(\alpha), b(x)$	[mm]:	Local material width in the roll gap.
Δb	[mm]:	Width change, difference between entry and exit material width in a rolling pass: $\Delta b = b_0 - b_1$. (2.1)
d	[mm]:	Roll diameter.
h_0	[mm]:	Entry height of the rolled material.
h_1	[mm]:	Exit height of the rolled material.
h_m	[mm]:	Average height $h_m = 0.5 \cdot (h_0 + h_1)$ (2.2)
$h(\alpha), h(x)$	[mm]:	Local height in the roll gap. $h(\alpha) = h_1 + 2 \cdot r \cdot (1 - \cos \alpha)$ (2.3) for $0 \leq \alpha \leq \alpha_0$, $h(x) = h_1 + 2 \cdot r \cdot \left(1 - \sqrt{1 - \frac{x^2}{r^2}}\right)$ (2.4) for $0 \leq x \leq l_d$. Approximation by Taylor series: $\cos \alpha \approx 1 - \frac{\alpha^2}{2}$ and $\sqrt{1 - \frac{x^2}{r^2}} \approx 1 - \frac{x^2}{2 \cdot r^2}$, $h(\alpha) \approx h_1 + r \cdot \alpha^2$, (2.5) $h(x) \approx h_1 + \frac{x^2}{r}$ (2.6)

Δh	[mm]:	Height deformation, height reduction. Difference between entry and exit height during a pass: $\Delta h = h_0 - h_1$ $\Delta h = 2 \cdot r \cdot (1 - \cos \alpha_0) \approx r \cdot \alpha_0^2$. (2.7)
l_0	[mm]:	Starting length.
l_1	[mm]:	Final length.
l_d	[mm]:	Roll bite length, length of the roll bite in rolling direction, horizontal projection: $l_d = \sqrt{r \cdot \Delta h - \dfrac{\Delta h^2}{4}} \approx \sqrt{r \cdot \Delta h}$. (2.8) The approximation $\sqrt{r \cdot \Delta h}$ corresponds with the chord $\overline{EA} = c$. Assuming biting angles $\alpha_0 \leq 20°$ the error is $\leq +1{,}5\%$.
r	[mm]:	Roll radius.
n	[min^{-1}]:	Roll revolution.
v_0	[m s^{-1}]:	Entry speed of the material.
v_1	[ms^{-1}]:	Exit speed of the material.
v_U	[m s^{-1}]:	Roll peripheral speed.
x	[mm]:	Horizontal roll gap coordinate: It is: $0 \leq x \leq l_d$.
$\dfrac{b_0}{h_0}$	[-]:	Width-to-height ratio.
$\dfrac{r}{h_1}$	[-]:	Height ratio, number for roll gap geometry.

$\dfrac{l_d}{h_m}$, $\dfrac{l_d}{h_0}$	[-]:	Roll gap ratio, number for roll gap geometry. Some rolling technique numbers like efficiency of deformation or deformation restance are described as function of the roll gap ratio. Further used geometric numbers are the height ratio and the related height reduction. The relation between $\dfrac{l_d}{h_m}, \dfrac{r}{h_1}$ and the related height reduction is: $$\dfrac{l_d}{h_m} = 2 \cdot \sqrt{\dfrac{r}{h_1}} \cdot \dfrac{\sqrt{\varepsilon_h \cdot (1-\varepsilon_h)}}{(2-\varepsilon_h)}. \qquad (2.9)$$ In case of constant ε_h, $\dfrac{l_d}{h_m}$ is proportional $\sqrt{\dfrac{r}{h_1}}$.
A_0	[mm²]:	Entry area of the rolled material.
A_1	[mm²]:	Exit area of the rolled material.
A_d	[mm²]:	Roll bite area, projection of the contact area between rolled material and roll onto the horizontal area.

$\overline{A\text{-}A}$	:	Exit section of the rolled material.
$\overline{E\text{-}E}$	:	Entry section of the rolled material.
$E-A$	:	Arc length of the roll in the roll gap [rad]: $E\text{-}A = \alpha_0 \cdot r$.
V_0	[mm³]:	Volume at roll gap gap entry.
V_1	[mm³]:	Volume at roll gap gap exit.
V	[mm³]:	Volume in the roll gap. Due to mass conservation and constant material density the law of volume constancy is valid: $V_0 = V_1 = V = \text{const.}$ (2.10)
α	[°, rad]:	Rolling angle (angle coordinate) in the roll gap: It is: $0 \leq \alpha \leq \alpha_0$. The rolling angle corresponds with roll gap coordinate x.
α_0	[°, rad]:	Biting angle between the roll radius vectors of the entry and exit section.
α_{0max}	[°, rad]:	Maximum possible biting angle which just fulfills the biting condition $(\tan \alpha_{0max} = \mu)$.
α_N	[°, rad]:	Neutral angle.

Relative width deformation

ε_b [-]: $\varepsilon_b = \dfrac{b_0 - b_1}{b_0} = 1 - \dfrac{b_1}{b_0}$. (2.11)

Relative height deformation

ε_h [-]: $\varepsilon_h = \dfrac{h_0 - h_1}{h_0} = 1 - \dfrac{h_1}{h_0}$. (2.12)

Very often ε_h is given in percentage (%).

Average related height reduction. By integration over the contact length the can be calculated according to:

ε_{hm} [-]: $\varepsilon_{hm} = \dfrac{1}{l_d} \cdot \int_0^{l_d} \varepsilon \, dx =$

$1 - \dfrac{1}{l_d} \int_0^{l_d} \dfrac{h(x)}{h_0} dx = \dfrac{2}{3} \cdot \varepsilon_h$ (2.13)

β [-]: Spreading ratio $\beta = \dfrac{b_1}{b_0} \geq 1$. (2.14)

γ [-]: Compression ratio $\gamma = \dfrac{h_1}{h_0} < 1$. (2.15)

Length ratio $\lambda = \dfrac{l_1}{l_0} > 1$. (2.16)

Due to the law of volume constancy it can be written:

λ [-]: $V_0 = b_0 \cdot h_0 \cdot l_0 = b_1 \cdot h_1 \cdot l_1 = V_1$. Therefore it is:

$\dfrac{V_1}{V_0} = \dfrac{b_1}{b_0} \cdot \dfrac{h_1}{h_0} \cdot \dfrac{l_1}{l_0} = \beta \cdot \gamma \cdot \lambda = 1$. (2.17)

Using $A_0 = b_0 \cdot h_0$ and $A_1 = b_1 \cdot h_1$ provides
$$V_0 = A_0 \cdot l_0 = A_1 \cdot l_1 = V_1$$
and $\lambda = \dfrac{l_1}{l_0} = \dfrac{A_0}{A_1}.$ \hfill (2.18)

Local deformation, related infinitesimal deformation:

$d\varphi_b$ [-]:   In width direction db: $d\varphi_b = \dfrac{db}{b}.$ \hfill (2.19)

$d\varphi_h$ [-]:   In height direction dh: $d\varphi_h = \dfrac{dh}{h}.$ \hfill (2.20)

Logarithmic deformation:

$d\varphi_l$ [-]:   In length direction dl: $d\varphi_l = \dfrac{dl}{l}$ \hfill (2.21)

φ_b [-]:   In width direction: $\varphi_b = \displaystyle\int_{b_0}^{b_1} \dfrac{db}{b} = \ln\dfrac{b_1}{b_0}$ \hfill (2.22)

φ_h [-]:   In height direction: $\varphi_h = \displaystyle\int_{h_0}^{h_1} \dfrac{dh}{h} = \ln\dfrac{h_1}{h_0}$ \hfill (2.23)

Between the logarithmic deformation and the related height deformation the relation

$$\varphi_h = \ln\dfrac{h_1}{h_0} = \ln\left(\dfrac{h_0 \cdot (1-\varepsilon_h)}{h_0}\right) = \ln(1-\varepsilon_h) \hfill (2.24)$$

exists.

Average logarithmic deformation:

φ_{hm} [-]: In height direction $\varphi_{hm} = \ln\left(1 - \frac{2}{3} \cdot \varepsilon_h\right)$ (2.25)

φ_{hl} [-]: In length direction $\varphi_l = \int_{l_0}^{l_1} \frac{dl}{l} = \ln\frac{l_1}{l_0}$ (2.26)

The law of volume constancy states: $\frac{V_1}{V_0} = \frac{b_1}{b_0} \cdot \frac{h_1}{h_0} \cdot \frac{l_1}{l_0} = 1$.

It follows: The sum of all three logarithmic deformations in width, height and length direction is zero:

$\ln\frac{V_1}{V_0} = \ln(1) = \ln\left(\frac{b_1}{b_0} \cdot \frac{h_1}{h_0} \cdot \frac{l_1}{l_0}\right) = \ln\frac{b_1}{b_0} + \ln\frac{h_1}{h_0} + \ln\frac{l_1}{l_0} = 0$ or

$\varphi_b + \varphi_h + \varphi_l = 0$ (2.27)

Pass (Rolling pass): One passage of the rolled material through a roll bite.

Roll gap: Area between the rolls where deformation takes place. It starts at the entry section $\overline{E\text{-}E}$ and ends at the exit section $\overline{A\text{-}A}$. According to the elementary theory this area is identical with the deformation zone despite the boundary areas of the deformation zone are not rectangular to each other.

3. Biting and rolling condition

a.) Biting condition

The possible thickness reduction in each rolling pass is limited by the biting angle and the friction conditions in the roll gap. During the biting process the rolled material has to be gripped first by the rolls and then pulled through the roll bite. In other words: The forces tearing the material into the roll bite must be higher than the forces pushing the material back. This efforts special rolling conditions which result under assuming Coulomb friction from the elementary mechanic force laws. The biting and rolling conditions are important when rolling heavy plates while the cold rolling process of strip material can be supported by the acting strip tensions.

The normal force F_N acting from the rolls onto the rolled material causes the tangential effective friction force $F_{Rb} = \mu \cdot F_N$, **Fig. 3.1**. Without this acting friction force no rolling is possible. While the horizontal component of the friction force $F_{Rb} \cdot \cos \alpha_0$ aims to pull the rolled material into the roll bite, the horizontal component of the normal force $F_N \cdot \sin \alpha_0$ aims the opposite. The value of both force components determines the biting condition, i.e. the relation

$$F_{Rb} \cdot \cos \alpha_0 = \mu \cdot F_N \cdot \cos \alpha_0 > F_N \cdot \sin \alpha_0 \qquad (3.1)$$

or

$$\mu > \tan \alpha_0 \approx \alpha_0 \qquad (3.2)$$

has to be fulfilled. The approximation $\tan\alpha_0 \cong \alpha_0$ is tolerable due to biting angles for strip rolling in the finishing train in magnitude of $\leq 10°$.

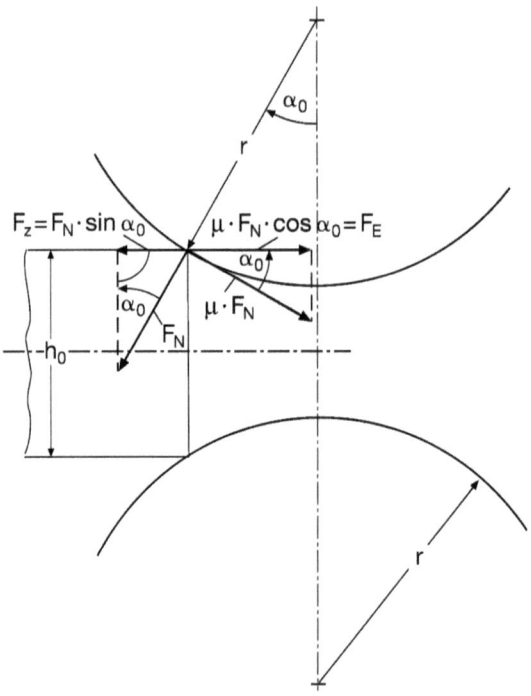

Fig. 3.1: Acting forces at the contact area between work rolls and rolled material during biting process according to /9/

Fulfilling the biting condition means the biting angle α_0 has to be less than the friction angle γ_R with $\tan\gamma_R = \mu$. The biting condition for the normal flat rolling process using work rolls with an identical radius follows according to

$$0 < \tan\alpha_0 < \mu. \tag{3.3}$$

Using the nomenclature given in Fig. 3.1 the biting angle α_0 can be written as

$$\cos \alpha_0 = 1 - \frac{h_0 - h_1}{2 \cdot r}. \qquad (3.4)$$

Knowing of the friction value µ and especially its influencing parameters is of essential importance for determination the maximum biting angle and therefore the biting conditions for the roll pass.

Main influencing parameters on the friction value during hot rolling are:

- Surface condition and chemical composition of the rolls and the rolled material,
- Temperature of the rolled material,
- Roll peripheral speed,
- Spezific roll force and
- Relative movement between the rolls and the rolled material.

A well-known equation for calculation the friction value in hot rolling was supplied by Geleji. This equation takes the influence of rolled material temperature T as well as rolls peripheral speed v_U and rolled material itself into account. For ground steel and chilled cast iron rolls this equation is:

$$\mu = 0.82 - 0.0005 \cdot T - 0.056 \cdot v_U. \qquad (3.5)$$

This formula can be used only if the minimum friction value is in between 0.2 and 0.25. If the lower value is achieved there is no significant influence of increasing roll peripheral speed on the friction value.

For preventing strip slippers or strip bumpings and for fulfilling the biting conditions several measures are possible in practice.

According to (3.4) the biting angle can be influenced by decreasing the absolute thickness reduction and increasing the work roll diameter. The friction value μ can be increased by roughening the work roll surface or by decreasing rolling temperature and or rolling speed according to (3.5).

From the practical point of view one has to consider the limit values for rolling force and rolling torque in the complete finishing train if load shifting is done by reducing absolute thickness reduction at the first finishing stand.

Roughened work rolls can cause rolled in scale on the strip surface.

When decreasing rolling temperatures and rolling speed the fulfillment of the final rolling temperature - especially when rolling thin gage hot strip in the austenitic temperature region - has to be considered.

The consequence is: The strip head end part has to be shaped mechanically in the way that the biting conditions are fulfilled during the pass. Principally this can be done in two ways:

- Forming the strip head end by a press that it will be gripped by the work rolls during the pass (disadvantage: High investment).
- Forming the strip head end with a curved shape by using curved knives in the crop shear at the entry section of the finishing train. The kinetic energy of the incoming transfer bar causes a further thickness reduction of the curved shaped head end when bumping to the work rolls (disadvantage: Crop shear with two types of scissors for head and tail end; advantage: Lower investment).

b.) Rolling condition

In case the roll gap is completely filled with material the situation differs completely. Under the assumption that the normal force F_N acts under the angle $\alpha = \alpha_0/2$ in the center of the contact arc in the roll bite the equilibirium of the horizontal forces can be written as:

$$\mu \cdot F_N \cdot \cos \frac{\alpha_0}{2} = F_N \cdot \sin \frac{\alpha_0}{2} \qquad (3.6)$$

or

$$\mu = \tan \frac{\alpha_0}{2} \approx \frac{\alpha_0}{2}. \qquad (3.7)$$

In case the rolling process is initiated the pull through condition or rolling condition is:

$$0 < \tan \alpha_0 < 2 \cdot \mu. \qquad (3.8)$$

If the biting condition is not fulfilled rolling is nevertheless possible in case the rolling condition $\tan \alpha_0 < 2 \cdot \mu$ is given. This is only possible if the biting behavior is supported by external forces like forward tensions or the pass is diminished (lowering rolling speed or short-time roll gap opening and adjusting after successful pass begin). On the other hand the friction coefficient can be increased by the use of modified work roll grades and roughening the rolls.

The magnitude of the biting angles during rolling of strip and plate material is summarized in **Table 3.1**.

Rolling, surface condition and lubrication	α_{0max} [°]	Friction value μ [-]
Hot rolling		
Heavy plate	15 - 22	0.27 - 0.4
Hot wide strip, finishing train	4 - 12	< 0.25
Narrow strip	22 - 24	0.40 - 0.45
Cold rolling		
Clean ground rolls, mineral oil lubrication	3 - 4	0.05 - 0.07
Chromium steel rolls, not perfect ground, mineral oil lubrication	6 - 7	0.10 - 0.12

Table 3.1: Biting angles and friction values for rolling of strip and plate material

4. Kinematics of the roll gap

In this part the characteristic values and parameters featuring the velocity behavior of the rolled material during deformation are defined and explained.

When passing the roll gap the material is accelerated due to the reduction of its cross section from the entry speed v_0 to the exit speed v_1. The relation is given by the continuity equation

$$b_0 \cdot h_0 \cdot v_0 = b_1 \cdot h_1 \cdot v_1 \tag{4.1}$$

or using the material cross section $A = b \cdot h$

$$A_0 \cdot v_0 = A_1 \cdot v_1. \tag{4.2}$$

In plane deformation without spreading ($b_0 = b_1$), **Fig. 4.1**, the speed of the rolled material in the roll gap is:

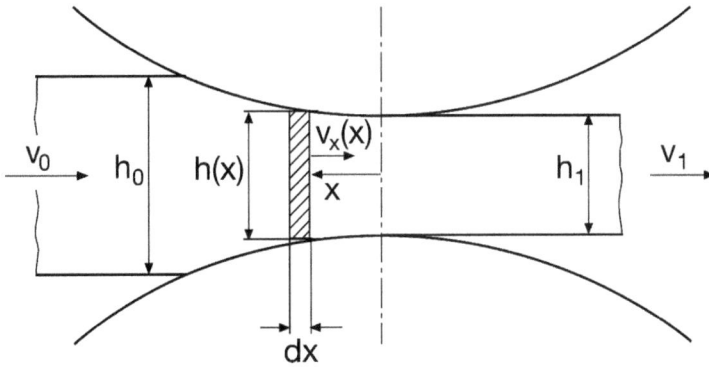

Fig. 4.1: Material speed in the roll gap during plane deformation /1/

$$h_0 \cdot v_0 = h(x) \cdot v(x) = h_1 \cdot v_1$$

$$v(x) = \frac{h_0}{h(x)} \cdot v_0 \qquad (4.3)$$

$$v(x) = \frac{h_1}{h(x)} \cdot v_1$$

Using $h(x)$ according to (2.6) it follows

$$v(x) = \frac{h_0}{h_1 + \frac{x^2}{r}} \cdot v_0 = \frac{h_1}{h_1 + \frac{x^2}{r}} \cdot v_1 \text{ with } 0 \leq x \leq l_d \qquad (4.4)$$

or with $h(\alpha)$ according to (2.5)

$$v(\alpha) = \frac{h_0}{h_1 + r \cdot \alpha^2} \cdot v_0 = \frac{h_1}{h_1 + r \cdot \alpha^2} \cdot v_1 \text{ with } 0 \leq \alpha \leq \alpha_0. \qquad (4.4a)$$

While the material speed in the roll gap increases from v_0 to v_1 the component of the roll peripheral speed in rolling direction $v_{ux} = v_u \cdot \cos \alpha$ remains almost constant (for $0 \leq \alpha \leq 20°$ the value for $\cos \alpha$ becomes $1 \geq \alpha \geq 0.94$). Therefore especially for strip rolling with rolling angles α_0 essential less than 20° it is:

$$v_{ux} \approx v_u \qquad (4.5)$$

Example cold rolling: d = 470 mm, h_0 = 4 mm, h_1 = 2.8 mm

$$\cos \alpha_0 = \frac{r - \frac{\Delta h}{2}}{r}$$

$$\cos \alpha_0 = \frac{235 - (4 - 2.8)/2}{235}$$

$$\cos \alpha_0 = 0.997 \approx 1$$

For plate rolling this approach is not regular due to larger rolling angles.

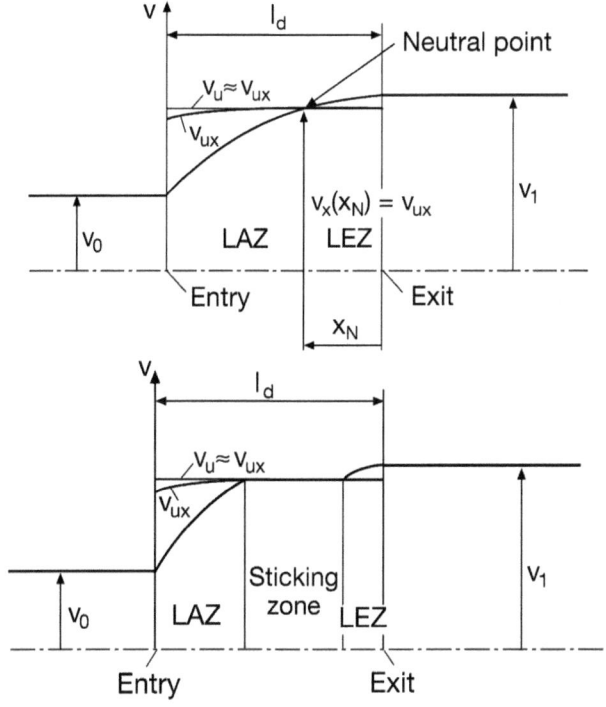

Fig. 4.2: Velocities of roll and rolled material /1/

Due to the force equilibrium it results by a given roll peripherical speed $v_0 < v_u$ and $v_1 > v_u$ if no forces, e.g. applied strip tension forces, act exterior of the roll gap on the rolled material.

Only at the neutral point ($x = x_N$ or $\alpha = \alpha_N$) the material speed and roll peripheral speed are equal, **Fig. 4.2** top. The rolled material speed from roll bite entry to the neutral point is smaller than the roll peripheral speed; therefore it is named lag zone LAZ. Between neutral point and roll gap exit the rolled material speed exceeds the roll peripheral speed; therefore this area is named forward zone LEZ.

Large friction especially in hot rolling can cause a static friction zone instead of the neutral point, **Fig. 4.2** bottom.

In the real rolling process the material flow is more or less inhomogeneous; this means the material speed v is not constant in height direction h. Assuming a constant velocity profile at the entry and exit of the roll gap and enlarging the neutral point to a static friction zone in which no relative movement of the rolls and rolled material takes place a concave velocity profile is formed between the roll gap entry and the neutral point; a convex velocity profile is built up between the neutral point and roll gap exit. The extension of the static friction zone being in between the two slipping zones is enlarged when friction increases, **Fig. 4.3**.

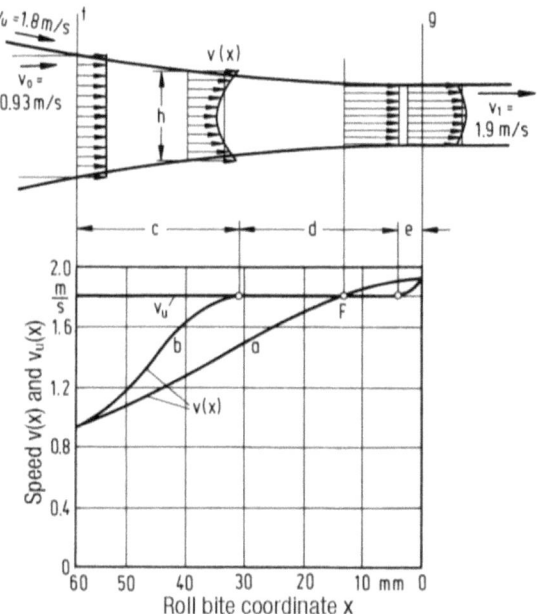

Fig. 4.3: Speed distribution in the roll gap when rolling hot strip /9/

The position of the neutral point x_N is given by the equilibrium of all horizontal forces acting on the rolled material within and outside the roll bite gap. For rolling without forces in length direction (no applied tension) the position of the neutral point is given according to the elementary theory. Siebel postulates the equation:

$$x_N = \frac{l_d}{2} \cdot \left(1 - \frac{l_d}{2 \cdot \mu \cdot r}\right) \tag{4.6}$$

or with $l_d \approx r \cdot \alpha_0$

$$x_N = \frac{r \cdot \alpha_0}{2} \cdot \left(1 - \frac{\alpha_0}{2 \cdot \mu}\right). \tag{4.6a}$$

This equation states that in rolling cases with large rolling angles and less friction ($\alpha_0/2 \cdot \mu \approx 1$) the neutral point is near the roll gap exit ($x_N \approx 0$). For $\alpha_0 = 2 \cdot \mu$ the rolling condition according to (3.8) is just fulfilled. On the other hand the value $x_N = l_d/2$ can not be exceeded for small rolling angles and large friction values.

Due to the uncertainty when determing the friction values equation (4.6) allows only a rough estimation of the neutral point position. Further external forces (strip tension) acting on the material cause a shifting of the neutral point, forward forces in direction to the roll bite entry and backward forces in direction the roll roll bite exit.

The magnitude of the slip forward depends on the neutral point position. The slip forward value corresponds with the speed difference at the roll bite exit: $v_1 - v_u = v_1 - (\pi \cdot d \cdot n)/60 = v_1 - (\pi \cdot r \cdot n)/30$. This parameter is of importance when determining the rotational speed steps for the rolling stands in a continuously operating rolling train where the material is rolled simultaneously in several stands. The magnitude of the existing tension between the rolling stands depends on the speed difference between the strip exit speed and the strip entry speed in the next stand. Due to speed control of the rolling stands the slip forward has to be considered in a continuous rolling process.

The expression (related) slip forward means:

$$\kappa = \frac{v_1 - v_u}{v_u} = \frac{v_1}{v_u} - 1 > 0, \qquad (4.7)$$

$$\kappa = \left(\frac{v_1}{v_u} - 1\right) \cdot 100\%. \qquad (4.7a)$$

The material height at the neutral point position is given by (2.5) and (2.6).

With $h_N/h_1 = v_N/v_1 = v_1/v_u$ (condition of continuity) equation (4.7) can be written as:

$$\kappa = \frac{r \cdot \alpha_N^2}{h_1} \qquad (4.8)$$

or

$$\kappa = \frac{h_1 + x_N^2/r}{h_1} = 1 + \frac{x_N^2}{r \cdot h_1}. \qquad (4.8a)$$

Examples for the neutral point position and slip forward:

<u>1.) Hot rolling of a slab</u>
$h_0 = 230\,mm;\ h_1 = 190\,mm;\ r = 400\,mm;\ \mu = 0.4$

$$x_N' = \frac{x_N}{l_d} = \frac{1}{2} \cdot \left(1 - \frac{\alpha_0}{2 \cdot \mu}\right) = \frac{1}{2} \cdot \left(1 - \frac{\sqrt{\Delta h/r}}{2 \cdot \mu}\right) = 0.30$$

$$\kappa = \frac{\Delta h}{h_1} \cdot x_N'^2 = 0.019 \triangleq 1.9\%$$

2.) Hot rolling of a strip
$h_0 = 40$ mm; $h_1 = 21$ mm; $r = 390$ mm; $\mu = 0.2$
$x_N' = 0.224$; $\kappa = 0.045 \triangleq 4.5\%$

3.) Cold rolling of a strip
$h_0 = 1$ mm; $h_1 = 0.7$ mm; $r = 230$ mm; $\mu = 0.05$
$x_N' = 0.32$; $\kappa = 0.043 \triangleq 4.3\%$

In case the material speed at the mill stand exit is larger than at the entry of the next mill stand the material will be compressed or a material loop is formed. During hot strip rolling in normal case a small loop is formed between the stands and controlled by loopers. In cold rolling strip tension is acting between the stands.

The rolling mill stand exit speed of the strip is:

$$v_1 = (1+\kappa) \cdot v_u. \qquad (4.9)$$

The condition of continuity in (4.2) can be expressed using the strip section $A = b \cdot h$:

$$A_1 \cdot (1+\kappa_1) \cdot v_{u_1} = A_2 \cdot (1+\kappa_2) \cdot v_{u_2} = A \cdot (1+\kappa) \cdot v_u. \qquad (4.10)$$

Denoting v_1 as exit speed and v_0 as entry speed into the roll bite rolling under tension takes place if $v_{1_1} < v_{0_2}$ (e.g. between the first and second stand) is valid. The corresponding speed difference is in between 1% and 3%. Loop formation between the stands assumes $v_{1_1} > v_{2_2}$.

When an infinitesimal narrow strip with the height h is compressed by dh in the time increment dt without spreading in the roll gap the compression speed is dh/dt. For calculation of the deformation rate the deformation is to be differentiated according to time:

$$\dot{\varphi}_h = \frac{d\varphi_h}{dt} = \frac{dh}{h \cdot dt} = \frac{1}{h} \cdot \frac{dh}{d\alpha} \cdot \frac{d\alpha}{dt}. \qquad (4.11)$$

Together with (2.5) it can be written:

$$\frac{dh}{d\alpha} = 2 \cdot r \cdot \alpha. \qquad (4.12)$$

With the angular velocity ω

$$\omega = \frac{d\alpha}{dt} \qquad (4.13)$$

and $v_u = \omega \cdot r$ it follows for $\dot{\varphi}_h$:

$$\dot{\varphi}_h = \frac{1}{h} \cdot 2 \cdot v_u \cdot \alpha. \qquad (4.14)$$

Substituting h by (2.5) provides:

$$\dot{\varphi}_h = 2 \cdot v_u \cdot \frac{\alpha}{(h_1 + r \cdot \alpha^2)} \qquad (4.15)$$

Using the mean value theorem the average deformation rate is expressed according to:

$$\dot{\varphi}_m = \frac{1}{\alpha_0} \cdot \int_0^{\alpha_0} \dot{\varphi}_h \cdot d\alpha = \frac{2 \cdot v_u}{\alpha_0} \cdot \int_0^{\alpha_0} \frac{\alpha}{(h_1 + r \cdot \alpha^2)} \cdot d\alpha \ . \tag{4.16}$$

Integration and simplifying finally leads to:

$$\dot{\varphi}_m = \frac{v_u}{l_d} \cdot \varphi_h \ . \tag{4.17}$$

In many cases the exit speed v_1 of the material is used for calculation. The material speed in the roll gap is:

$$v = \frac{dx}{dt} \ . \tag{4.18}$$

Using (2.6) provides

$$\frac{dh}{dx} = 2 \cdot x \cdot r \ . \tag{4.19}$$

The deformation rate can be expressed as

$$\dot{\varphi}_h = \frac{dh}{h \cdot dt} = \frac{1}{h} \cdot \frac{dh}{dx} \cdot \frac{dx}{dt} = \frac{1}{h} \cdot \frac{2 \cdot x}{r} \cdot v \tag{4.20}$$

or together with (2.6) and (4.3)

$$\dot{\varphi}_h = \frac{2 \cdot v_1 \cdot h_1}{r} \cdot \frac{x}{\left(h_1 + \frac{x^2}{r}\right)^2} \ . \tag{4.21}$$

The average deformation rate

$$\dot{\varphi}_m = \frac{1}{l_d} \cdot \int_0^{l_d} \dot{\varphi} \cdot dx \qquad (4.22)$$

can be written according to Weber /9/

$$\dot{\varphi}_m = \frac{v_1}{l_d} \cdot \varepsilon_h . \qquad (4.23)$$

Examples for the magnitude of the deformation rate:

1.) Hot rolling of a slab
$h_0 = 240 \, mm$; $h_1 = 190 \, mm$; $r = 400 \, mm$; $v_u = 2.5 \, m\,s^{-1}$

$$\dot{\varphi}_m = \frac{v_u}{l_d} \cdot \varphi_h \approx \frac{v_u}{\sqrt{r \cdot \Delta h}} \cdot \ln\left(\frac{h_0}{h_1}\right)$$

$$\dot{\varphi}_m = \frac{2.5 \, m\,s^{-1}}{\sqrt{400 \, mm \cdot (240 \, mm - 190 \, mm)}} \cdot \ln\left(\frac{240 \, mm}{190 \, mm}\right) = 4.1 s^{-1}$$

2.) Hot rolling of a strip
$h_0 = 3 \, mm$; $h_1 = 2.6 \, mm$; $r = 300 \, mm$; $v_u = 7 \, m\,s^{-1}$

$$\dot{\varphi}_m = 91.4 \, s^{-1}$$

3.) Cold rolling of a strip
$h_0 = 1 \, mm$; $h_1 = 0.65 \, mm$; $r = 200 \, mm$;
$v_u = 1200 \, m\,min^{-1} = 20 \, m\,s^{-1}$

$$\dot{\varphi}_m = 1029.8 \, s^{-1}$$

5. Application of the volume constancy law in flat rolling process

The law of volume constancy in the rolling process in general is given by (2.17):

In rolling processes material lengths are difficult to determine; therefore the length ratios λ are combined with the cross sections. Due to (2.18) $\lambda = A_0/A_1$ can be written as:

$$A_0 = A_1 \cdot \lambda \text{ or } A_1 = A_0/\lambda, \qquad (5.1)$$

if the first pass is calculated in a sequence of rolling passes. For the following passes analogue:

$$\frac{A_1}{A_2} = \lambda_2 \text{ or } A_1 = \lambda \cdot A_2. \qquad (5.2)$$

In general it is:

$$A_{n-1} = A_n \cdot \lambda_n. \qquad (5.3)$$

n is denoting the number of the pass. For an arbitrary number of passes n:

$$\frac{A_0}{A_n} = \lambda_{ges} = \lambda_1 \cdot \lambda_2 \cdot \ldots \cdot \lambda_{n-1} \cdot \lambda_n = \prod_{i=1}^{n} \lambda_i \qquad (5.4)$$

Another equation when considering volume constancy is of importance in strip mills. The volume flow rate $\dot{V} = A \cdot v$ is as well constant in the complete mill (condition of continuity):

$$\dot{V} = A \cdot v = A_0 \cdot v_0 = A_1 \cdot v_1 = A_n \cdot v_n \qquad (5.5)$$

This means the total length ratio λ_{ges} is identical to the ratio between final rolling speed and entry speed into the first mill stand:

$$\frac{A_0}{A_n} = \lambda_{ges} = \frac{v_n}{v_0}. \qquad (5.6)$$

The final rolling speed differs due to the rolled material grade, the exit material section and the kind of rolling mill. In hot wide strip mills the final rolling speed is in between 8 m/s and 25 m/s. But not only final rolling speed has to be considered; as well material entry speed v_0 in the first mill stand is of importance. This speed has to be chosen high enough to avoid excessive cooling down of the bar in front of the finishing mill with the consequence of crack formation. Furthermore low entry speeds increase the high thermal loads on the work rolls causing increased roll wear and roll cracks. When rolling hot strip the entry speed in the finishing train is normally between 0.4 m/s and 1 m/s.

Moreover, the total reduction (bar-strip) is an important parameter influencing speed limit in combination with the final temperature to be reached and affected by the layout of the mill.

In the following an example of transfer bar thickness to be calculated for a strip with the final dimensions 2.0 mm x 1200 mm and the final rolling speed $v_{End} = 9$ m s^{-1} is given:

$$\lambda_{ges} = \frac{v_{End}}{v_0} = \frac{9}{0.4} = 22.5,$$

$$A_0 = A_n \cdot \lambda_{ges} = (2\,\text{mm} \cdot 1200\,\text{mm}) \cdot 22.5 = 54{,}000\,\text{mm}^2,$$

$$\Rightarrow h_0 = 45\,\text{mm with a strip width of } b = 1200\,\text{mm}.$$

The second example gives the calculation of the final rolling speed for a hot strip with the final dimension 1.5 mm x 1000 mm and – as further parameters – the entry height $h_0 = 35$ mm and mill entry speed $v_0 = 0.6$ m/s:

$$v_{End} = \frac{A_0}{A_n} \cdot v_0$$

$$v_{End} = \lambda_{ges} \cdot v_0$$

$$v_{End} = \frac{1000\,\text{mm} \cdot 35\,\text{mm}}{1000\,\text{mm} \cdot 1.5\,\text{mm}} \cdot 0.6\,\text{m s}^{-1}$$

$$v_{End} = 14\,\text{m s}^{-1}$$

5.1 Calculation of production in strip rolling

For continuous rolling of a hot strip in the finishing train the mill productivity P can be calculated according to:

$$P = \frac{m_{sl} \cdot a_f}{t_r + t_p}\,[\text{kg s}^{-1}]. \tag{5.7}$$

In practice the mill productivity is given in $[t\ h^{-1}]$, therewith:

$$P = \frac{m_{sl} \cdot a_f}{t_r + t_p} \cdot \frac{3{,}600}{1{,}000} \quad [t\ h^{-1}] \qquad (5.8)$$

with

m_{sl}	[kg]:	Slab weight,
a_f	[-]:	Yield factor, $a_f \leq 1$,
t_r	[s]:	Rolling time,
t_p	[s]:	Idle time between two strips (slabs, bars).

The rolling time t_{ri} of the strip in a mill stand i generally is:

$$t_{ri} = \frac{l_i}{v_i} \qquad (5.9)$$

with:

l_i	[m]:	Rolling length in stand i,
v_i	[m s^{-1}]:	Strip speed in stand i.

Due to the volume constancy condition it follows:

$$l_i = \frac{V}{A_i} = \frac{m_{sl} \cdot a_f}{\rho \cdot A_i} \qquad (5.10)$$

where ρ denotes the material density. For steel grades $\rho = 7.55 \cdot 10^{-6}$ kg mm^{-3} is assumed in general.

Summarizing (5.8) to (5.10) gives for the mill productivity:

$$P = \frac{m_{sl} \cdot a_f \cdot 3{,}6}{\frac{m_{sl} \cdot a_f}{\rho \cdot A_i \cdot v_i} + t_p} \quad [t\ h^{-1}], \tag{5.11}$$

under the assumption that the rolling speed remains constant.

In hot strip rolling speed up mode is practiced for the compensation of temperature losses in the transfer bar in front of the finishing mill. It is distinguished between

- Normal temperature speed up: $a_{su} \approx 0.04 \div 0.08\ m\ s^{-2}$,
- Increased temperature speed up: $a_{su} \approx 0.1 \div 0.3\ m\ s^{-2}$,
- Power speed up $a_{su} \approx 0.5 \div 1.2\ m\ s^{-2}$.

During rolling with speed up a continuously increasing strip speed is given with the acceleration a after the transit time t_{tr} of the strip head from the last active finishing mill stand to the coiler (starting strip coiling). The total strip rolling time then amounts to:

$$t_r = t_{tr} + t_a \tag{5.12}$$

with

t_{tr}	[s]:	Transit time,
t_a	[s]:	Acceleration time for temperature or power speed up.

The rolled strip length can be calculated according to the law of volume constancy:

$$l_{tot} = v_0 \cdot t_{tr} + v_0 \cdot t_a + \frac{a_{su} \cdot t_a^2}{2} . \quad (5.13)$$

t_a is the acceleration time for the corresponding speed up mode and v_0 the strip starting speed before strip acceleration begins.

For the complete strip length the equation is valid:

$$l_{tot} = \frac{m_{sl} \cdot a_f}{\rho \cdot b \cdot h_1} = \frac{s_{coil} \cdot b \cdot a_f}{\rho \cdot b \cdot h_1} = \frac{s_{coil} \cdot a_f}{\rho \cdot h_1} \quad (5.14)$$

using the parameters

b [mm]: Hot strip width,
h_1 [mm]: Hot strip thickness,
s_{coil} [kg mm^{-1}]: Spezific coil weight.

Defining l_{tr} as rolled strip length during transit time and l_{Rest} as rolled strip during acceleration provides the equation for the complete strip length l_{tot}:

$$l_{tot} = l_{tr} + l_{Rest} \quad (5.15)$$

or

$$l_{Rest} = l_{tot} - l_{tr} \quad (5.15a)$$

and together with (5.14)

$$I_{Rest} = \frac{S_{coil} \cdot a_f}{\rho \cdot h_1} - v_0 \cdot t_{tr} .$$ (5.16)

Using this equation the acceleration time t_a can be calculated to:

$$v_0 \cdot t_{tr} + \frac{a_{su} \cdot t_{tr}^2}{2} = I_{Rest} = \frac{S_{coil} \cdot a_f}{\rho \cdot h_1} - v_0 \cdot t_a .$$ (5.17)

This quadratic equation provides the solution for the acceleration time:

$$t_a = -\frac{v_0}{a_{su}} + \sqrt{\left(\frac{v_0}{a_{su}}\right)^2 + \frac{2}{a_{su}} \cdot \left(\frac{S_{coil} \cdot a_f}{\rho \cdot h_1} - v_0 \cdot t_{tr}\right)} .$$ (5.18)

Knowing the acceleration time t_a the complete strip rolling time can be calculated using (5.12).

Using the general formula work for uniformly accelerated movement several equations for strip rolling with speed up can be derived if the final rolling speed is known.

$$t_{su} = \frac{v_{End} - v_0}{a_{su}} , \quad a_{su} = \frac{v_{End} - v_0}{t_{su}} ,$$
$$a_{su} = \frac{v_{End}^2 - v_0^2}{2 \cdot I_{Rest}} , \quad I_{Rest} = \frac{v_{End}^2 - v_0^2}{2 \cdot a_{su}} .$$ (5.19)

In the same way the time calculation can be performed if additional acceleration phases, e.g. power speed up, take place.

5.2 Calculation of production in reversing rolling

For calculation of mill productivity in case of reversing rolling, e.g. slab, heavy plate or reversing cold rolling, it is:

$$P = \frac{m_{sl} \cdot a_f \cdot 3.6}{t_{tot}} \ [t\ h^{-1}]. \tag{5.20}$$

The total rolling time t_{tot} follows according to

$$t_{tot} = \sum_{i=1}^{n}(t_{Ri} + t_{Pi}) + t_f \tag{5.21}$$

using

t_{Ri} [s]: Rolling time of an individual pass i,
t_{Pi} [s]: Idle time between the passes,
t_f [s]: Aftermath times for consecutive slabs or strips.

For the single pass times in the reversing mill stand it is:

$$t_{Ri} = \frac{l_i}{v_i} = \frac{l_0 \cdot \sum_{j=1}^{i}\lambda_i}{v_i}. \tag{5.22}$$

(5.22) is an averaged calculation formula because reversing mill stands operate with accelerating mill drives producing a complicated time sequence during a reversing pass.

6. Spreading

In contrast to the assumption of plane deformation in the elementary stripe model a part of the material reduced from its height flows not into length but into width. Material spreading is enlarged the smaller the ratio width per height is. Further influencing parameters on spreading can be summarized as follows:

Increasing
- h/b-ratio raises spreading,
- l_d/b-ratio raises spreading,
- related height reduction $|\varepsilon_h|$ raises spreading,
- roll radius r raises spreading,
- friction value μ raises spreading,
- yield strength k_f raises spreading,
- temperature T lowers spreading and
- rolling speed v lowers spreading.

The relationships are very complex; therefore it is not possible to predict the spreading behavior using elementary methods. Furthermore spreading is not constant in height direction; a width profile is formed. A reliable determination of material spreading during deformation is only possible using numerical methods.

For the estimation of averaged material spread in material height direction some empirical formulas had been developed describing the influencing parameters by power- and exponential functions. Some are listed in **Table 6.1**.

Spreading equation according to	
Geuze	$b_1 = b_0 + C \cdot \Delta h$; $C=0.35$ for steel
Tafel and Sedlaczek	$b_1 = b_0 + 0.17 \cdot \Delta h \cdot \sqrt{r/\Delta h}$
Siebel	$b_1 = b_0 + C \cdot \Delta h \cdot l_d/h_0$; $C=0.35$ for steel
Bachtinow and Schternow	$b_1 = b_0 + 0.58 \cdot \Delta h \cdot (l_d - \Delta h/2 \cdot \mu)/h_0$
Hill	$b_1 = b_0 \cdot (h_0/h_1)^W$; $W = 0.5 \cdot \exp(-b_0/2 \cdot l_d)$
Wusatowski	$b_1 = b_0 \cdot (h_0/h_1)^W$; $W = 10^{-1.27 \cdot (b_0/h_0) \cdot (h_0/2 \cdot r)^{0.56}}$
Sander	$b_1 = b_0 \cdot (h_0/h_1)^W$, $W = 10^{-0.76 \cdot (b_0/h_0)^{0.39} \cdot (b_0/l_d)^{0.12} \cdot (h_0/r)^{0.59}}$
Pawelski	$b_1 = b_0 \cdot \left[1 + B_1 \cdot \left(\dfrac{\Delta h}{h_0}\right)^{B_2} \cdot \left(\dfrac{l_d}{h_0}\right)^{B_3} \cdot \left(\dfrac{h_0}{b_0}\right)^{B_4}\right]$; B_1 to B_4 are constants

Table 6.1: Empiric spreading formula /1, 6/

In the formula according to Geuze width increase is described as a material dependent part of height reduction (steel: C = 0.35). All other parameters are not considered. Therefore Geuze's formula is only a rough estimation of material spreading.

The spreading formulas according to Tafel and Sedlaczek as well as Bachtinov and Schternow are further developments of the Geuze formula. In these formulas further parameters like r/h_0, l_d/h_0-and $\Delta h/h_0$ are taken into account.

In the equations according to Hill, Wusatowski and Sander the parameters are considered due to an exponential approach. The influence of relative material width (b_0/l_d, b_0/h_0) is taken into account in the equations. This can be understood considering that the material flow is hindered at the contact areas of the rolls. The material flow is in direction of the lower flow hindering region, i.e. in direction of the lower dimension (l_d, b) of the contact area. The larger the ratio b/l_d is the lower the material spreading will be.

The more general spreading formula according to Pawelski gives the advantage that some spreading formulas can be derived from this formula; e.g. the Geuze formula with $B_1 = 0.35$, $B_2 = B_4 = 1$ and $B_3 = 0$.

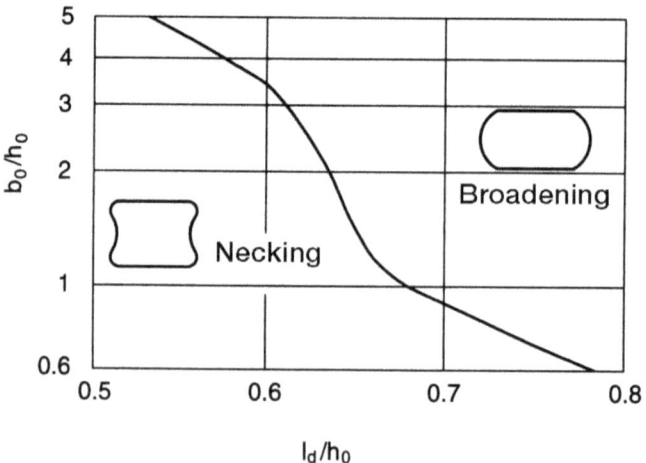

Fig. 6.1: Side formation during hot flat rolling depending on roll gap geometry according to Dahl /1/

In case of incoming rectangular cross sections width varation is not constant regarding the material height. Two typical types of cross section contours are observed after rolling. Depending on the roll gap and side ratio the width profile change from a convex to a concave shape, **Fig. 6.1**.

The calculated final width b_1 denotes the average value over the material height, **Fig. 6.2**.

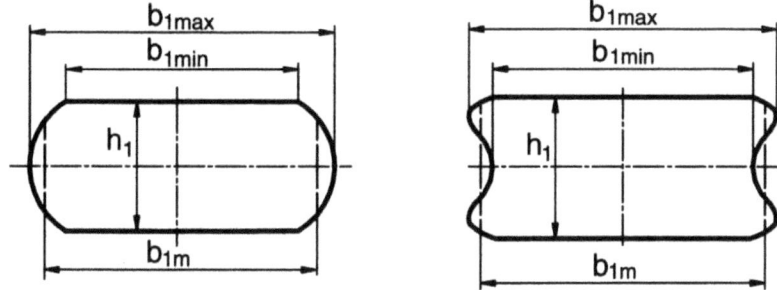

Fig. 6.2: Width profile during hot flat rolling /1/

Thereby

$$b_{1m} \equiv b_1 = \frac{A_1}{h_1} \approx \frac{2 \cdot b_{1max} + b_{1min}}{3} \tag{6.1}$$

represents a commodity and

$$b_1 = \frac{A_1}{h_1} \approx \frac{b_{1max} + b_{1min}}{2} \tag{6.2}$$

a rough approximation.

7. Roll flattening and minimum strip thickness during cold rolling

Caused by the pressure from the strip onto the roll elastic deformation of the rolls takes place. The original circular contour of the rolls changes in the contact area into a curve with an increased radius, **Fig. 7.1**.

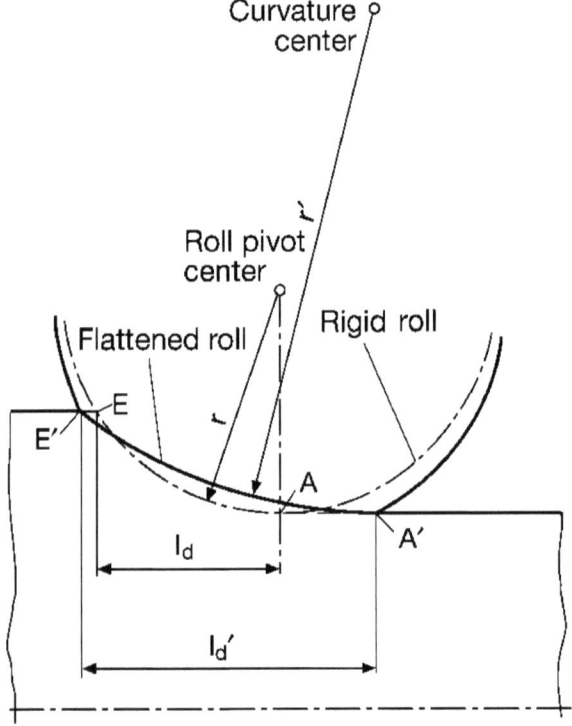

Fig. 7.1: Elastic roll deformation (roll flattening) according to /1/

Hitchcock proved according to the Hertz theory that the roll contour in the contact area has a constant curvature radius $r' > r$. In the roll gap an elliptic stress distribution is assumed in Hitchcock's theory. This assumption had been proved by measuring the stress distribution in the roll gap. Hitchcock noticed the advantage for calculating the contact length l_d' in the roll gap by substituting the roll radius r by r' in the contact area. For $r' \leq 2 \cdot r$ the curvature radius in the flattened area the Hitchcock equation can be applied.

$$r' = r \cdot \left(1 + C_H \cdot \frac{F}{b \cdot \Delta h}\right), \qquad (7.1)$$

and therewith

$$l_d' = \sqrt{r' \cdot \Delta h}, \qquad (7.2)$$

with F as rolling force and C_H work roll material constant. C_H is named work roll flattening constant and is calculated by

$$C_H = \frac{16}{\pi} \cdot \frac{1 - v^2}{E}. \qquad (7.3)$$

In (7.3) v denotes the Poisson's number, E is the Young's modulus of the roll. For steel rolls with $v = 0.3$ and $E = 2.1 \cdot 10^5$ N mm^{-2} according to (7.3) C_H values to $C_H = 22 \cdot 10^{-6}$ mm^2 N^{-1}. It can be further deduced that rolls flatten more if the specific roll force F/b is increased and the flattening constant is enlarged. For constant rolling parameters and reduction of roll flattening roll material grades with high Young's moduli like ceramic grades are to be applied. Ceramic roll grades with the further advantage of improved wear resistance are actually under development for cold rolling applications and wire rod rolling mills.

The rolling force for a pass is calculated by an iteration method. Starting point for iteration is the non flattened roll radius. The resulting flattened radius is used for the next iteration step. The iteration circle is done as long as two calculated roll forces differ out of a limiting degree.

In case of hot rolling the rolling pressures are minor than in case of cold rolling. Therefore roll flattening often will be neglected when calculating the effective force if the average material flow stress is less than 500 to 600 N/mm². In this case the assumption $r = r'$ is made.

Cold rolling of especially very thin gage strip gives a rapid increase of flow stress due to increased friction with decreasing thickness. This results in extended roll flattening. Limit cases exist where plastic deformation is not possible and only elastic deformation of the rolls and the embedding of the strip into the rolls takes place exclusively. Therefore the knowledge of the minimum strip gage is necessary up to which rolling is possible under given rolling conditions.

Under the assumption that rolling of very thin gage strip material can be compared with the forming process between two parallel areas relations can be derived which illustrate up to which minimum strip thickness rolling is possible.

Following the Hertz and Hitchcock theory and assuming sliding friction in the roll gap with the friction coefficient µ and in limit case of occurring linear flattening with an infinite roll radius the lowest rollable strip thickness is approximately given by Stone /13/:

$$h_{min} = 1.545 \cdot \mu \cdot C_H \cdot k_f \cdot r. \tag{7.4}$$

The equation is valid for very small thickness reductions $\varepsilon_h \approx 0$ like skin pass rolling; k_f denotes the yield stress of the strip.

Taking into consideration that especially in the area of the neutral point in the roll gap sticking friction occurs the following equation can be used for a rough calculation of the minimum rollable thickness /14/:

$$h_{min} = 2 \cdot \mu \cdot C_H \cdot k_f \cdot r. \tag{7.5}$$

So, due to (7.4) and (7.5) rolling of thinner strips is possible reducing the friction coefficient, using softer steel grades, decreasing the roll diameter or applying more rigid roll grades. Therefore foil material with a final thickness of some 10 µm are processed on very stiff Multi Roller Stands like a 20-high Sendzimir or Roon configuration. The stiffness is necessary to avoid thickness deviation over the width of the strip by bending of the rolls. The work roll radii of a 20-high mill are in between 15 and 30 mm. The remaining nine rolls on each mill stand side are located semi circularly around the top and bottom work roll as support to fix their position.

The applied strip tensions influence the rollable minimum strip thickness as roll force decreases with increased strip tension. A simple approximation for the calculation of the minimum thickness following /15/ is

$$h_{min} = 3.58 \cdot \mu \cdot r \cdot \frac{(k_f - (\sigma_B + \sigma_F))}{E}. \qquad (7.6)$$

σ_B denotes the backward tension and σ_F the forward or coiler tension.

8. Flow curves of metallic materials

For the calculation of the integral parameters of rolling processes like roll force, roll torque and rolling power the knowledge of the material's yield stress at given rolling conditions is essential.

Up to the material's yield strength only elastic deformation occurs in the material; releasing the force the material reverses to its intial state. The situation differs above the yield strength: In this case a part of the deformation remains in the material.

The yield strength depends on the material and its structure condition and is usually determined by tensile testing. For solid materials the value is always greater zero. For calculation of the flow curve, the stress-strain-diagram can be used up to the ultimate tensile strength if corrections are made concerning true area of loading and calculating the logarithmic plastic deformation φ by $\varphi = \ln(1+\varepsilon_{pl})$ with ε_{pl}: plastic deformation.

By definition, the yield stress k_f is the amount of stress which causes the start of plastic material flowing under uniaxial stress conditions. The amount is introduced because the stress will be negative in case of compression stress conditions; so

$k_f = \sigma_1$ uniaxial tension stress, (8.1)

$k_f = |\sigma_1|$ uniaxial compression stress. (8.2)

In general, k_f is is a function of the grade, deformation φ, deformation rate $\dot\varphi$ and temperature T:

$$k_f = k_f\left(\text{Grade}, \varphi, \dot\varphi, T\right). \tag{8.3}$$

The knowledge of the flow curve is necessary concerning the material strain hardening during and after deformation.

At ambient room temperature metals strain hardening takes place when deformation increases. Under elevated temperatures strain hardening occurs, too, but will be superimposed by softening and finally recovery as well as recrystallization if the thermo-mechanical conditions allow this.

8.1 Parameters influencing the yield stress

Beside the grade (chemical composition) and the processing history (structure) yield stress depends on deformation φ, deformation rate $\dot\varphi$ and forming temperature T. Thereby not only the actual values for φ, $\dot\varphi$ and T are to be considered, but also the previous forming and temperature history influencing the (actual) structural conditions.

In a simplified manner in practice flow curves $k_f(\varphi)$ are determined for different values of $\dot\varphi$ and T; this procedure provides sufficient information regarding flow stress values

$k_f\left(\varphi, \dot{\varphi}, T\right)$ for the calculation of local parameters like stress and deformation in case of inhomogeneous deformation and temperature distribution, see **Fig. 8.1**.

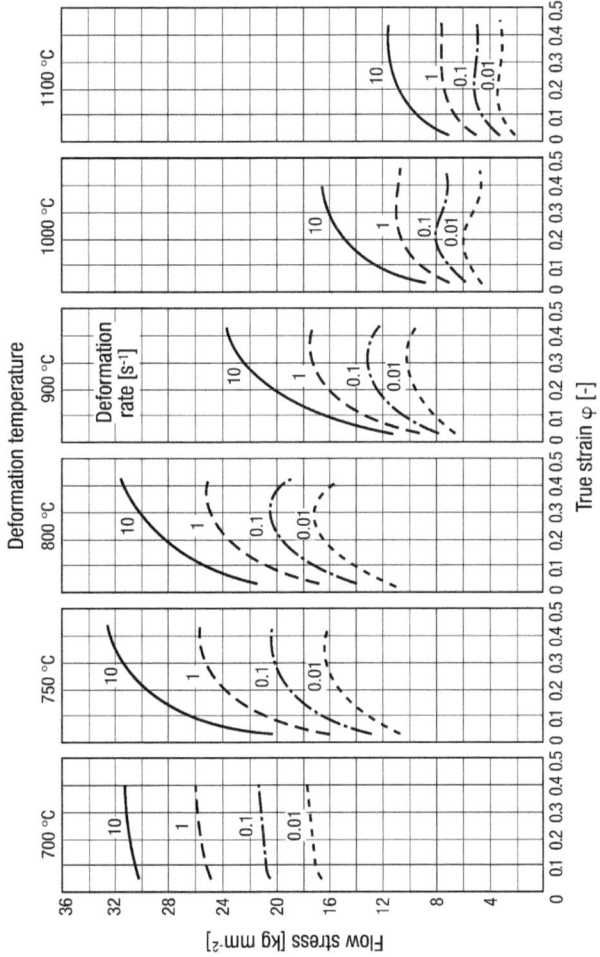

Fig. 8.1: Influence of deformation on the yield stress during the hot deformation for steel grade C45 /12/

Influence of the deformation

Cold flow curves

In general metallic grades strengthen at ambient room temperature with increasing deformation φ, caused by an increased dislocation density due to deformation. Exceptions are metals - such as lead - recrystallizing below room temperature. Thereby material softening takes place compensating strengthening; the consequence is an ideal plastic material behavior ($k_f \neq k_f(\varphi)$), if only focusing on φ.

Hot flow curves

During hot deformation strengthening and thermal activated softening processes (dynamic recovery and recrystallization) take place. This can cause an almost constant or a decrease of the flow curve to an even lower constant niveau under certain deformation conditions, **Fig. 8.1**.

Influence of the deformation rate

The influence of the deformation rate $\dot{\varphi}$ on the yield stress is mainly based on softening effects. While strengthening takes place parallel to deformation, softening goes on with a temperature dependent speed (thermal acivitated process). In case of high deformation rates the material has less time for softening compared to low deformation rates. In general, the yield stress increases with increasing deformation rates. Experimental results describing the influence of the deformation rate on the yield stress are reliable within the range $0 \leq \dot{\varphi} \leq 50 \ s^{-1}$.

Cold flow curves

In general softening is a thermally activitated process. Therefore the influence of the deformation rate on the yield stress is only small for most metallic grades at room temperature or temperatures below. As an approximation in the range $0 \leq \dot{\varphi} \leq 100 \text{ s}^{-1}$ it can be assumed, that flow stress increases by 3 - 10 % if the deformation rate increases by a decimal power.

Hot flow curves

In this case the influence of the deformation rate is higher than in the case of deformation at room temprerature. In the range of recrystallization temperature, which in general is roughly 2/3 of the material's melting temperature and aimed generally during hot deformation, the increase of flow stress is 50 to 80 % if the deformation rate is increased by a decimal power, **Fig. 8.2**.

Influence of the temperature

In the range of hot forming temperature T the flow curve is similar to an exponential function. Deviations from this tendency are given in the temperature range of blue brittleness or in the range of phase transformation, precipitations as well as for multi-phase materials. This dependency can not be described by only one function in the temperature range starting from room temperature and ending at temperatures above 1000 °C. Several functions valid for different temperature regions have to be applied.

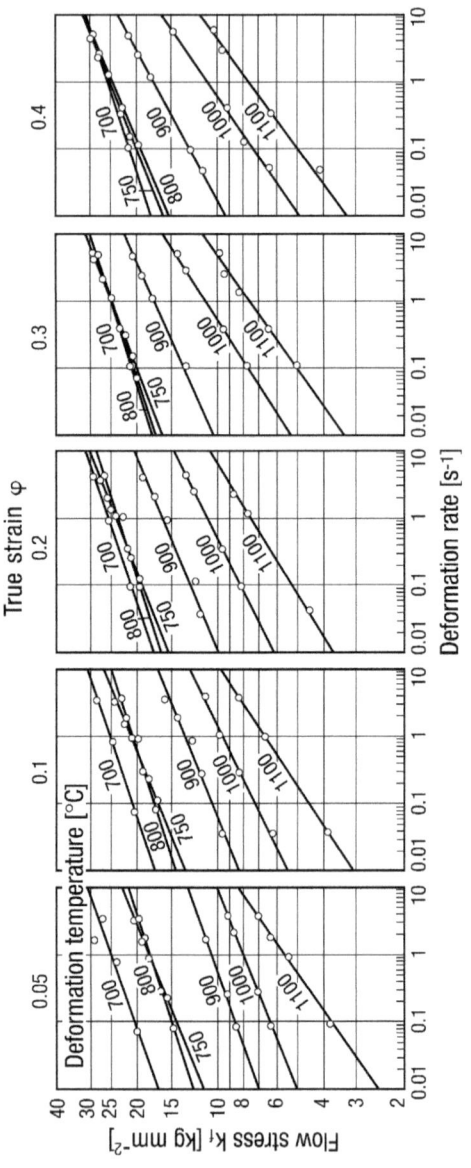

Fig. 8.2: Influence of the deformation rate on yield stress during hot deformation for steel grade C45 /12/

Multi-stage hot flow curves

In practice, the total reduction of the material is performed in several passes. For the calculation of the parameters per pass, like roll force and roll torque, the question arises how far the continuous flow curves can be used for determining the material flow stress.

In principal the qualitative trend of multi-stage and continuous flow curves is the same. The average flow stress k_{fm} is more or less identical. Therefore the same rules and relationships as for continuous flow curves can be applied when mathematically describing multi-stage flow curves.

Due to softening in the idle times between the passes the starting value of the flow stress partly decreases rapidly before the next deformation step; but due to early beginning restrengthening in the next pass the k_f-values of the origin flow curve are achieved more or less. The flow curve of the (i+1)-th stage can be interpreted as prosecution of the i-th stage.

The material flow stress depends only minor on the temporal sequence of deformation steps. The value is increased if in the pass sequence deformations are performed ranging in the region of the flow curve maximum and if the deformed material softens quickly or idle times and temperatures enable completely material softening. In this case the k_f-values for each pass are equal to the maximum of the flow curves.

In general in a continuous operating rolling mill the idle times between two passes allow no complete material softening. In this case the flow stress can be also calculated in analogy to the continuous flow curve; the deformations per pass have to be summed up under consideration of the softened material part. The remaining material strain hardening is to be taken into account in the hot deformation temperature range when calculating the flow stress especially for idle times /19/

< 2 s for unalloyed carbon steels and low alloyed steels and
< 3 s for austenitic and high alloyed tool steels.

This fact meets especially for the last stands of a hot strip finishing train.

8.2 Mathematical modeling of flow curves

The complex relationship between yield stress and processing history of a metal makes it difficult to describe the yield stress with a distinct function or determine a unique modeling for a certain parameter combination. Further uncertainties are given by the determination of the flow curves e.g. choice of the testing method, inhomogenities of stress- and deformation state, shifting of test parameters during the trial (temperature, deformation rate). These uncertainties have to be counteracted by a careful execution of the respective trials.

For many practical applications a simplified model describing the yield stress as a function of the actual metal temperature, deformation and deformation rate is applied. For estimation of rolling forces, rolling torque and work as well as rolling power using elementary calculculation methods is in general sufficient though superficial simulation tools nowadays deliver improved results.

The following models for mathematical description of the yield stress are proven for many alloys:

Cold flow curves

For describing cold flow curves often the equation

$$k_f = A_\varphi \cdot \varphi^n \qquad (8.4)$$

is used with $A_\varphi = k_f(\varphi = 1)$. n is the strengthening exponent. In double logarithmic scaling it is $\log k_f = \log A_\varphi + n \cdot \log \varphi$, a slope with the inclination n. The strengthening exponent is assumed to be constant regarding φ; this is only partly the case. Therefore extrapolation of the flow curves in deformation ranges not measured in tensile tests should not be done. Using this model effort for the complete description the knowledge of the initial yield stress $k_f(\varphi = 0) = k_{f0}$ is necessary.

Due to large errors in the range of very small deformations $0 \leq \varphi \leq 0.02$ the formula (8.4) is modified:

$$k_f = k_{f0} + A_{\varphi 1} \cdot \varphi^{n_1} \qquad (8.5)$$

or

$$k_f = A_{\varphi 2} \cdot (b + \varphi)^{n_2}. \qquad (8.6)$$

The influence of the deformation rate is very often not considered because of its only small effect on the yield stress especially at ambient room temperature. It can be modeled by an exponential function:

$$k_f \approx A_{\dot\varphi} \cdot \dot\varphi^m \tag{8.7}$$

with $A_{\dot\varphi} = k_f\left(\dot\varphi = 1\right)$. The velocity exponent m is in the range of $0.001 \leq m \leq 0.07$. It is significant smaller than in case of hot deformation. Together with formula (8.4) the complete model for cold flow curves can be written as:

$$k_f = A_\varphi \cdot \varphi^n \cdot A_{\dot\varphi} \cdot \dot\varphi^m \tag{8.8}$$

or

$$k_f = A \cdot \varphi^n \cdot \dot\varphi^m \tag{8.9}$$

The coefficients and exponents in (8.7) to (8.9) make necessary to use $\dot\varphi$ with the unit s^{-1}.

Hot flow curves

Due to its similarity with a power function the strengthening behavior of the flow curve $k_f(\varphi)$ can be modeled by

$$k_f \sim A_\varphi \cdot \varphi^n. \tag{8.10}$$

For softening caused by dynamic recovery and recrystallization formulas like

$$k_f \sim \exp(-m_\varphi \cdot \varphi) \tag{8.11}$$

or

$$k_f \sim \exp(m_\varphi/\varphi) \tag{8.12}$$

are often used for modeling hot flow curves.

The complete influence of the deformation on the yield stress can be described according to

$$k_f = A_\varphi \cdot \varphi^n \cdot \exp(-m_\varphi \cdot \varphi). \tag{8.13}$$

For describing the influence of the deformation rate as well a power function approach is chosen:

$$k_f \sim A_{\dot\varphi} \cdot \dot\varphi^{m_{\dot\varphi}}. \tag{8.14}$$

In this approach the velocity exponent $m_{\dot{\varphi}}$ is considered to be constant. In reality it increases with increasing $\dot{\varphi}$ and it is furthermore depending on temperature.

For temperatures T above the $(\alpha-\gamma)$ - transformation (A_3-temperature) the influence of yield stress is given by an exponential function

$$k_f \sim A_T \cdot \exp(-m_T \cdot T). \tag{8.15}$$

Summarizing (8.13) to (8.15) the approach for the hot flow curve is given by:

$$k_f = K \cdot A_\varphi \cdot \varphi^n \cdot \exp(-m_\varphi \cdot \varphi) \cdot A_{\dot{\varphi}} \cdot \dot{\varphi}^{m_{\dot{\varphi}}} \cdot A_T \cdot \exp(-m_T \cdot T). \tag{8.16}$$

The coefficients of the several flow curve formulas are determined by regression analysis. This assumes enough measuring values $k_f(\varphi, \dot{\varphi}, T)$ e.g. from tensile tests. For a hot flow curve in the parameter range between $0 \leq \varphi \leq 0.8$, $1 s^{-1} \leq \dot{\varphi} \leq 10 s^{-1}$ and $1000\,°C \leq T \leq 1200\,°C$ for minimum three temperatures and for each temperature minimum three deformation rates flow curves up to the deformation of $\varphi = 0.8$ are to be determined. The regression software needs roughly 20 to 50 pairs of values $k_f = f(\varphi)$ for each of the nine flow curves.

A mathematical flow description can be possibly adulterated despite the tensile tests had been carried carefully. Therefore flow curve functions should only be applied if all parameters had been checked in detail.

Despite the modeling of flow curves for several material grades there are only few flow curves available for the large number of steel materials rolled in hot and cold rolling mills. Nevertheless, there is a big effort to solve this problem using simulation models like FEM, FDM and cellular models.

Material grade	Temperature range	HSF
Spring steel	850 °C – 1150 °C	1.35 – 1.10
Austenitic stainless steel	900 °C – 1200 °C	1.90 – 1.60
High-alloyed stainless steel	900 °C – 1200 °C	2.20 – 1.80
Ferritic steels	800 °C – 1150 °C	0.80 – 0.90
Cold work steel	950 °C – 1100 °C	2.00 – 1.60
High-speed alloy steel	850 °C – 1150 °C	2.60 – 1.90
High-alloyed tool steels	900 °C – 1100 °C	2.80 – 2.00
Ni-basic alloys	900 °C – 1200 °C	3.80 – 2.50
Superalloys (highly heat resisting)	900 °C – 1300 °C	5.00 – 3.50

Table 8.1: Hard steel factors (HSF) for several steel grades

The measurement and evaluation of flow curves for all steel grades is time and cost intensive. Therefore as a first approach the quotient of the not known strength of a material grade to a grade which strength is known is estimated. The so called hard steel factor or steel factor (HSF) is calculated according to:

$$\frac{k_f\left(\text{spezific steel grade}, \varphi, \dot{\varphi}, T\right)}{k_f\left(C15, \varphi, \dot{\varphi}, T\right)} = HSF\left(\varphi, \dot{\varphi}, T\right) \qquad (8.17)$$

As reference the low carbon steel grade C15 is chosen. As an approximation value for $HSF\left(\varphi, \dot{\varphi}, T\right)$ the values regarding to **Table 8.1** are used.

These HSF-values can be applied as first approximation of the mill load to check wether the mill is overloaded or the load is within the acceptable limits. In case of limit dimensions or grades the exact determination of flow curves is recommended.

9. Application of flow curves and deformation energy

An application for using flow curves is the calculation of the metal forming energy for a deformation process.

As an example, compression of a cylinder by height dh gives the formula for the deformation energy:

$$dW = F \cdot dh = \sigma_1 \cdot A \cdot dh = k_f \cdot A \cdot dh \qquad (9.1)$$

The deformation volume V can be written using A as area and h as height of the cylinder according to

$$V = A \cdot h \qquad (9.2)$$

or

$$A = \frac{V}{h}. \qquad (9.3)$$

For the deformation follows:

$$dW = k_f \cdot \frac{V}{h} \cdot dh$$
$$dW = k_f \cdot V \cdot \frac{dh}{h}. \qquad (9.4)$$
$$dW = k_f \cdot V \cdot d\varphi$$

For the total deformation energy ($h_0 \rightarrow h_1$) the formula finally can be written as

$$W = V \cdot \left| \int_{h_0}^{h_1} k_f \cdot d\varphi \right| \approx V \cdot k_{fm} \cdot \left| \int_{h_0}^{h_1} \frac{dh}{h} \right|,$$

$$W = V \cdot k_{fm} \cdot \left| \ln\left(\frac{h_1}{h_0}\right) \right| = V \cdot k_{fm} \cdot |\varphi_h|$$

(9.5)

Relating the deformation energy onto the volume provides the spezific deformation energy or deformation energy density

$$\varpi = \frac{W}{V} = k_{fm} \cdot |\varphi_h|.$$

(9.6)

The average flow stress for a certain material grade is given using the flow curve $k_f(\varphi)$ according to

$$k_{fm} = \frac{1}{\varphi_h} \cdot \int_0^{\varphi_h} k_f(\varphi) d\varphi.$$

(9.7)

The minimum deformation energy in case of friction free deformation is called ideal deformation energy and is calculated with

$$W_{id} = V \cdot k_{fm} \cdot |\varphi_h|.$$

(9.8)

The real deformation energy including friction losses is named effective deformation energy W_{eff}. Using this terminology the deformation efficiency factor η_F can be defined as the ratio between the ideal and effective deformation energy:

$$\eta_F = \frac{W_{id}}{W_{eff}}. \qquad (9.9)$$

The deformation power is:

$$P = \frac{W_{def}}{t_{def}}. \qquad (9.10)$$

In this equation W_{def} denotes the deformation energy and t_{def} the deformation time. These energy considerations can be used for the calculation of the deformation heat:

$$W_{def} = V \cdot k_{fm} \cdot |\varphi_h| = m \cdot c_P \cdot \Delta T. \qquad (9.11)$$

Deformation and heat energy are physically equivalent, therefore it is:

$$V \cdot k_{fm} \cdot |\varphi_h| = m \cdot c_P \cdot \Delta T \qquad (9.12)$$

using m [kg] as mass of the deformed material, c_P [J kg^{-1} K^{-1}] as specific heat capacity of the material grade and ΔT [K] as temperature gain.

Denoting $\rho = m \cdot V^{-1}$ as material density it is

$$k_{fm} \cdot |\varphi_h| = \frac{m}{V} \cdot c_P \cdot \Delta T = \rho \cdot c_P \cdot \Delta T \qquad (9.13)$$

or

$$\Delta T = \frac{k_{fm} \cdot |\varphi_h|}{c_P \cdot \rho}. \qquad (9.14)$$

Example for cold rolling of steel grade S235JR:

$h_0 = 2\,mm$, $h_1 = 1.4\,mm$, $k_{fm} = 650\,\dfrac{N}{mm^2}$, $c_P = 500\,\dfrac{J}{kg\,K}$,

$\rho = 7.85 \cdot 10^{-6}\,\dfrac{kg}{mm^3}$

$|\varphi_h| = \left|\ln\left(\dfrac{h_1}{h_0}\right)\right| = \left|\ln\left(\dfrac{1.4}{2}\right)\right| \approx 0.36$

$\Delta T = \dfrac{k_{fm} \cdot |\varphi_h|}{c_P \cdot \rho} = \dfrac{650 \cdot 0.36}{500 \cdot 7.85 \cdot 10^{-9}}\,\dfrac{N}{mm^2}\,\dfrac{kg\,K}{N\,m}\,\dfrac{mm^3}{kg}\,\dfrac{m}{mm} \approx 60\,K$

S235JR is a low alloyed and heat treated structural steel. Its former name is St 37-2.

10. Elementary rolling equation

For the calculation of the stress distribution in the roll gap for a non spreading rolling process (plane deformation) an elementary piece with the thickness dx, the height h and the width b is cut from the roll gap volume. The equilibrium condition between the acting internal and external forces is formulated in x-direction. As an external force the normal stress σ_N acts on the elementary volume element. Its direction is not rectangular to the rolled material, but it forms the angle α with the ordinate y, **Fig. 10.1**.

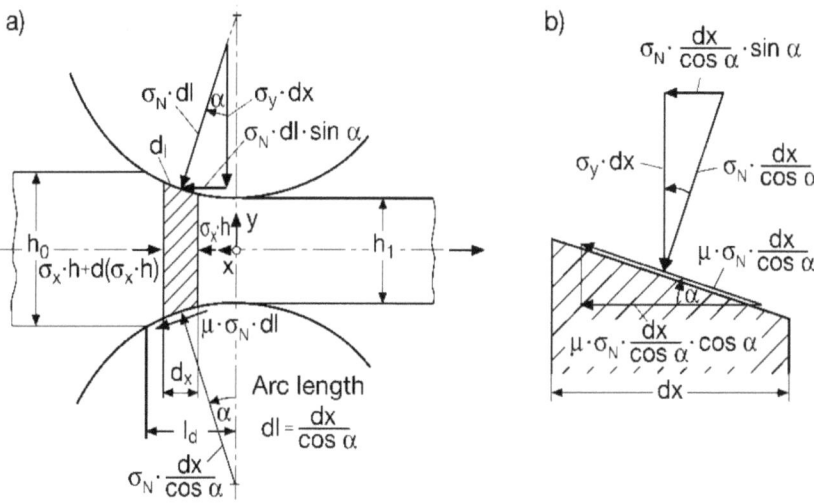

Fig. 10.1: Elementary rolling theory – Forces and stresses in the roll gap according to /1, 9/

$$F_{x1} = 2 \cdot \sigma_N \cdot \sin \alpha \cdot r \cdot d\alpha \cdot b = 2 \cdot \sigma_N \cdot \sin \alpha \cdot r \cdot b \cdot d\alpha . \quad (10.1)$$

The factor 2 arises due to the participation of top and bottom roll on the forming process, assuming symmetry.

The direction of this force is the positive x-coordinate and therefore its algebraic sign is positive. As further external force acts the frictional force between roll and material. This causes an external shear stress τ_R on the surface of the rolled material. To define the direction of this shear stress the lead and the lag zone shown in fig. 4.2 in the roll gap has to be distinguished. This was not necessary when giving the direction of the before deducted force F_{x1}. In the lag zone this force is directed to the point of origin. Therefore the horizontal component of the shear stress is given by:

$$F_{x2} = -2 \cdot \tau_R \cdot \cos \alpha \cdot r \cdot d\alpha \cdot b = -2 \cdot \tau_R \cdot \cos \alpha \cdot r \cdot b \cdot d\alpha. \quad (10.2)$$

The negative algebraic sign means that this force acts in negative x-axis, so in the exit direction. Elsewhere no further external forces are acting in x-direction if rolling is performed without tension on the strip between the stands. The horizontal stress σ_x decreases in the interval between x and x+dx by the amount $d\sigma_x$ because $\frac{d\sigma_x}{dx}$ has to be negative. The internal horizontal stress on the elementary material element - i.e. σ_x and $\sigma_x + d\sigma_x$ - acts as reaction stresses. The resulting internal force F_{x3} caused by these stresses can be written as:

$$F_{x3} = -d\sigma_x \cdot h \cdot b. \quad (10.3)$$

The external forces are in equilibrium with the internal forces; they can be equalized by $\sum_{i=1}^{3} F_{xi} = 0$. For the lag zone it follows:

$$-d\sigma_x \cdot h \cdot b + 2 \cdot \sigma_N \cdot \cos \alpha \cdot r \cdot b \cdot d\alpha - 2 \cdot \tau_R \cdot r \cdot b \cdot d\alpha = 0. \quad (10.4)$$

Simplifying and rearranging gives:

$$\frac{d(\sigma_x \cdot h)}{d\alpha} = 2 \cdot (\sigma_N \cdot \sin \alpha \cdot r - \tau_R \cdot \cos \alpha \cdot r). \quad (10.5)$$

Taking into consideration that in the lead zone only the algebraic sign of the force F_{x2} is different to the lag zone the elementary rolling equation for the lead zone follows according to:

$$-d\sigma_x \cdot h \cdot b + 2 \cdot \sigma_N \cdot \sin \alpha \cdot r \cdot b \cdot d\alpha + 2 \cdot \tau_R \cdot \cos \alpha \cdot r \cdot b \cdot d\alpha = 0 \quad (10.6)$$

or

$$\frac{d(\sigma_x \cdot h)}{d\alpha} = 2 \cdot (\sigma_N \cdot \sin \alpha \cdot r + \tau_R \cdot \cos \alpha \cdot r) \quad (10.7)$$

and therewith the elementary rolling equation for the complete roll gap is:

$$\frac{d(\sigma_x \cdot h)}{d\alpha} = 2 \cdot (\sigma_N \cdot \sin \alpha \cdot r \mp \tau_R \cdot \cos \alpha \cdot r). \quad (10.8)$$

The negative algebraic sign describes the lag zone, the positive one the lead zone.

According to the respective equilibrium condition between the vertical external forces and the interior compressive stress the resulting vertical stress is given by

$$\sigma_y = \sigma_N \pm \tau_R \cdot \tan \alpha. \qquad (10.9)$$

In this equation the positive algebraic sign denotes the lag zone and the negative sign the lead zone.

Transforming (10.8) on the roll gap coordinate $x = r \cdot \sin \alpha$ gives together with $dx/d\alpha = r \cdot \cos \alpha$:

$$\frac{d(\sigma_x \cdot h)}{dx} = 2 \cdot (\sigma_N \cdot \tan \alpha \mp \tau_R). \qquad (10.10)$$

(10.8) and (10.10) are ordinary differential equations with the three unknown functions σ_x, σ_N and τ_R. For solution, two further equations are necessary. In the following the mostly used rolling theories are described and it will be shown that all theories can be derived from the elementary rolling equation.

10.1 Elementary rolling equation according to Siebel

The works according to Siebel /18/ can be regarded in their complete form as the eldest solution of the rolling theory. The formulation of the stress distribution in the roll gap is done by Siebel assuming:

- A parallelepipedic deformation is considered in the roll gap. This means the vertical and horizontal stresses acting on a stripe element remain constant within the section and the former plane sections remain plane during rolling.
- A complete sliding region, that means no friction region between the rolls and the rolled material, is assumed. The friction coefficient is considered to be constant along the arc length. Furthermore the external shear stress τ_R is approximated by the product of normal stress σ_N and the friction coefficient μ, so

$$\tau_R \approx \mu \cdot \sigma_N . \tag{10.11}$$

- Due to small rolling angles Siebel equalized the normal stress and vertical stress in direction of the ordinate:

$$\sigma_N = \sigma_y . \tag{10.12}$$

- To solve (10.8) a further simplification was made: The vertical stress σ_y is assumed to be constant in the complete roll gap and equal to the yield stress k_f; In other words: Tresca's shear stress hypothesis states: $\sigma_y - \sigma_x = k_f$. Together with $\sigma_x \ll \sigma_y$ it can be notified: $\sigma_y \approx k_f$.

Using these assumptions (10.8) can be written as:

$$\frac{d(\sigma_x \cdot h)}{d\alpha} = 2 \cdot k_f \cdot r \cdot \sin \alpha \mp 2 \cdot k_f \cdot \mu \cdot r \cdot \cos \alpha , \tag{10.13}$$

Due to the assumptions made below the solution of (10.8) is limited:

- For small rolling angles α one can approximate:
  $\sin\alpha \approx \tan\alpha \approx \alpha$ und $\cos\alpha \approx 1$. (10.14)

- For small rolling angles α the arc element $r \cdot d\alpha$ can be equalized with the thickness dx of the volume element. It follows:
  $r \cdot d\alpha \approx dx$. (10.15)

- Roll flattening is not considered, i.e. the working roll remains unchanged in diameter.

Using these approximations (10.8) can be written as:

$$\frac{d(\sigma_x \cdot h)}{dx} = 2 \cdot k_f \cdot (\tan\alpha \mp \mu). \qquad (10.16)$$

The further use of $\tan\alpha = \frac{x}{r}$ provides the Siebel differential equation for the horizontal stress distribution in the roll gap:

$$\frac{d(\sigma_x \cdot h)}{dx} = 2 \cdot k_f \cdot \left(\frac{x}{r} \mp \mu\right). \qquad (10.17)$$

Integration gives:

$$\sigma_x \cdot h = 2 \cdot k_f \cdot \int_0^{l_d} \left(\frac{x}{r} \mp \mu\right) \cdot dx. \qquad (10.18)$$

Therefore the horizontal stress distribution σ_x is due to

$$\sigma_x = \frac{2 \cdot k_f}{h} \cdot \left(\frac{x^2}{2 \cdot r} \mp \mu \cdot x + C_1 \right). \tag{10.19}$$

The integration constant C_1 for tension free rolling in the lead zone follows according to the boundary condition $\sigma_x(x=0) = 0$.

Inserting in (10.19): $0 = \frac{2 \cdot k_f}{h} \cdot \left(\frac{0^2}{2 \cdot r} \mp \mu \cdot 0 + C_1 \right)$, this means $C_1 = 0$.

In analogy the boundary condition for the lag zone is formulated: $\sigma_x(x = l_d) = 0$, therefore it is:

$$0 = \frac{2 \cdot k_f}{h} \cdot \left(\frac{l_d^2}{2 \cdot r} - \mu \cdot l_d + C_1 \right) \Leftrightarrow C_1 = \mu \cdot l_d - l_d^2 / 2 \cdot r.$$

The horizontal stress distributions in the lead and lag zones are:

$$\sigma_{xLEZ} = \frac{2 \cdot k_f}{h_1 + x^2/r} \cdot \left(x^2/2 \cdot r + \mu \cdot x \right), \tag{10.20}$$

(Lead zone, $0 \leq x \leq x_N$),

$$\sigma_{xLAZ} = \frac{2 \cdot k_f}{h_1 + x^2/r} \cdot \left(x^2/2 \cdot r - \mu \cdot x + \mu \cdot l_d - l_d^2/2 \cdot r\right), \qquad (10.21)$$

(Lag zone, $x_N \leq x \leq l_d$).

The neutral point can be determined using the equilibrium condition $\sigma_{xLEZ} = \sigma_{xLAZ}$:

$$x_N = \frac{l_d}{2} \cdot \left(1 - \frac{l_d^2}{2 \cdot r \cdot \mu}\right). \qquad (10.22)$$

Transformation to the angle coordinate $\sin\alpha \approx \alpha = x/r$ gives the equations:

$$\sigma_{xLEZ} = \frac{2 \cdot k_f}{h_1 + r \cdot \alpha^2} \cdot \left(r \cdot \alpha^2/2 + \mu \cdot r \cdot \alpha\right) \qquad (10.23)$$

(Lead zone, $0 \leq \alpha \leq \alpha_N$),

$$\sigma_{xLAZ} = \frac{2 \cdot k_f}{h_1 + r \cdot \alpha^2} \cdot \left(r \cdot \alpha^2/2 - \mu \cdot r \cdot \alpha + \mu \cdot r \cdot \alpha_0 - r \cdot \alpha_0^2/2\right) \qquad (10.24)$$

(Lag zone, $\alpha_N \leq \alpha \leq \alpha_0$),

$$\alpha_N = \frac{\alpha_0}{2} \cdot \left(1 - \frac{r \cdot \alpha_0^2}{2 \cdot \mu}\right) \text{ (Neutral point)}. \qquad (10.25)$$

10.2 Elementary rolling equation according to von Kárman

The rolling theory according to Theodore von Kárman /16/ is quite similar to the solution of Siebel under the following simplifications and conditions:

- Homogeneous compression behavior without material spreading.
- The relationship between external shear stress τ_R and normal stress σ_N is given in analogy to Siebel by $\tau_R = \mu \cdot \sigma_N$.
- The yield stress k_f is considered to be constant along the roll bite width and therefore no material hardening takes place.
- Roll flattening is not taken into consideration.
- For small rolling angles α the approaches $\cos \alpha \approx 1$, $\sin \alpha \approx \alpha = \dfrac{x}{r}$ and $dx = r \cdot d\alpha$ are made. For small angles α the difference between normal and vertical stress can be neglected; i.e. the assumption $\sigma_N \approx \sigma_y$ is valid.
- The arc of circle of the rolls is approximately replaced by the parabolic function:
$$h(\alpha) = h_1 + 2 \cdot r \cdot (1 - \cos \alpha) \approx h_1 + x^2 / r = h(x). \qquad (10.26)$$

Using these approximations (10.8) can be written as:

$$\frac{d(\sigma_x \cdot h)}{dx} = \frac{d\sigma_x}{dx} \cdot h + \sigma_x \cdot \frac{dh}{dx} = 2 \cdot \sigma_y \cdot \left(\frac{x}{r} \mp \mu\right). \qquad (10.27)$$

According to (10.26) for the differential quotient dh/dx can be written:

$$\frac{dh}{dx} = 2 \cdot x \cdot r. \tag{10.28}$$

Therewith:

$$h \cdot \frac{d\sigma_x}{dx} + \sigma_x \cdot \frac{2 \cdot x}{r} = 2 \cdot \sigma_y \cdot \left(\frac{x}{r} \mp \mu\right). \tag{10.29}$$

As a second relation between the unknown functions $\sigma_x = f_1(x)$ and $\sigma_y = f_2(x)$ von Kárman introduced the yield criterion according to the maximum shear stress theory developed by Tresca. This theory postulates that the yield stress is equal to the difference between the major and minor principal stresses. Therewith the following relation is valid:

$$\sigma_y - \sigma_x = k_f. \tag{10.30}$$

Putting this relation into (10.29) and simplifying gives the von Kárman'sche differential equation for flat rolling:

$$\frac{d\sigma_x}{dx} \pm \frac{2 \cdot \mu}{h_1 + x^2/r} \cdot \sigma_x - \frac{2 \cdot k_f}{h_1 + x^2/r} \cdot \left(\frac{x}{r} \mp \mu\right) = 0. \tag{10.31}$$

For simplification the above differential equation is written as follows:

$$\frac{d\sigma_x}{dx} + f(x) \cdot \sigma_x + g(x) = 0. \tag{10.32}$$

This is an inhomogeneous linear differential equation with $f(x)$ and $g(x)$ as known continuously differentiable functions of the roll gap coordinate x.

For a given value $\sigma_x = \sigma_{x0}$ at a local position $x = x_0$ in the roll bite as boundary condition, the horizontal stress distribution can be written as

$$\sigma_x = \exp\left(\int_{x_0}^{x} f(x')\cdot dx'\right)\cdot\left[\sigma_{x0} - \int_{x_0}^{x} g(x'')\cdot \exp\left(\int_{x_0}^{x} f(x')\cdot dx'\right)\cdot dx''\right] \quad (10.33)$$

This solution is of practical importance if

- the functions f(x) and g(x) are formed simply and
- the integral can be evaluated mathematically closed.

10.3 Spezific pressure distribution in the roll gap

The pressure distribution in the roll gap according to the elementary rolling theories is shown in **Fig. 10.2** on the left hand side. The pressure characteristic in length direction is identical with the yield stress. From roll bite entry to the neutral point the stress increases to its maximum and decreases afterwards in the direction of the roll bite exit.

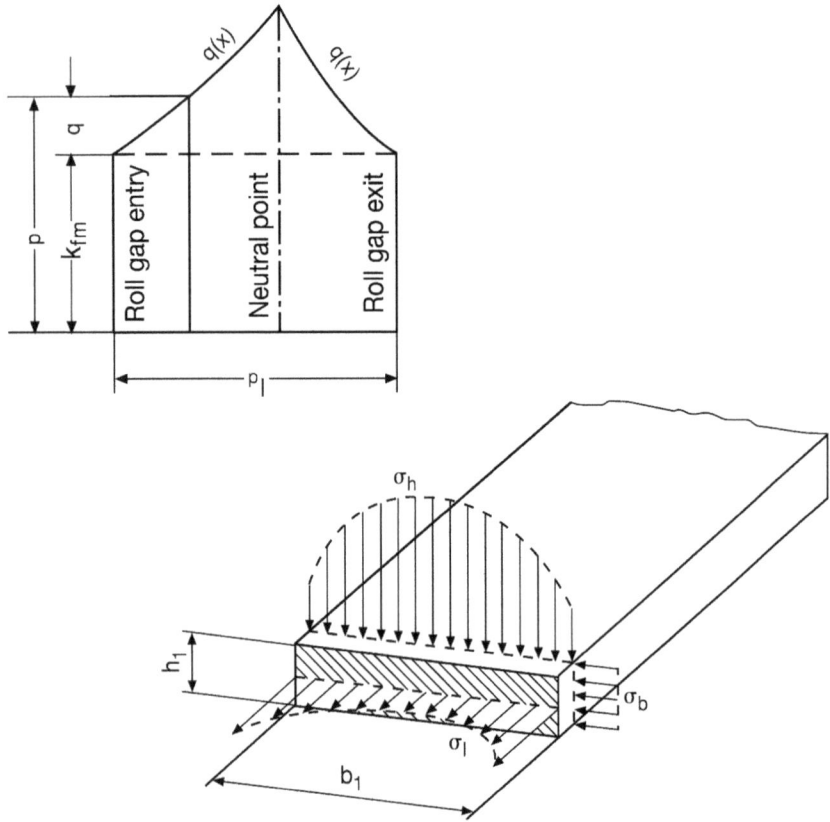

Fig. 10.2: Pressure p, q(x) (top) and stress σ (bottom) distribution in the roll gap due to elementary rolling theory /9/

Applying an exterior uniform distributed longitudinal stress on the material cross section gives a stress distribution in the roll gap as indicated on the right hand side in Fig. 10.2.

When external forces act at the roll gap the integration constant of the von Karman differential equation becomes different to zero. The curve segments are shifted due to the amount of the forward and backward tension, **Fig. 10.3**. According to the von Mises yield criterion it can be written for the roll gap entry

$$p_0 = 1.155 \cdot k_{f0} - \sigma_B \tag{10.45}$$

and for the roll gap exit

$$p_1 = 1.155 \cdot k_{f1} - \sigma_F . \tag{10.46}$$

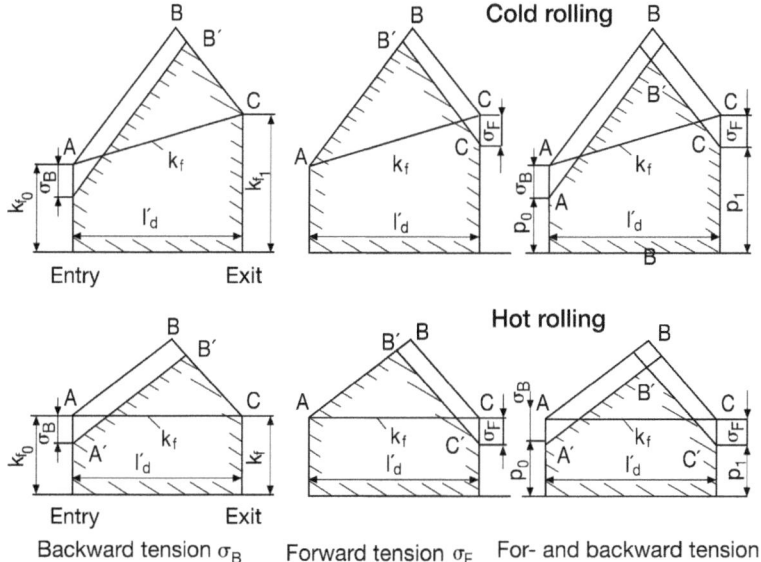

Fig. 10.3: Influence of strip tensions on the pressure distribution in the roll gap according to /9/

Due to strip tensions the rolling pressures at the roll gap entry and/or exit are decreased and the neutral point is shifted towards the opposite direction of the applied strip tension. This means forward tension shifts the neutral point in direction of roll gap entry and backward tension causes the opposite. In case the position of the neutral point is at the roll gap entry or exit the rolls start slipping on the strip.

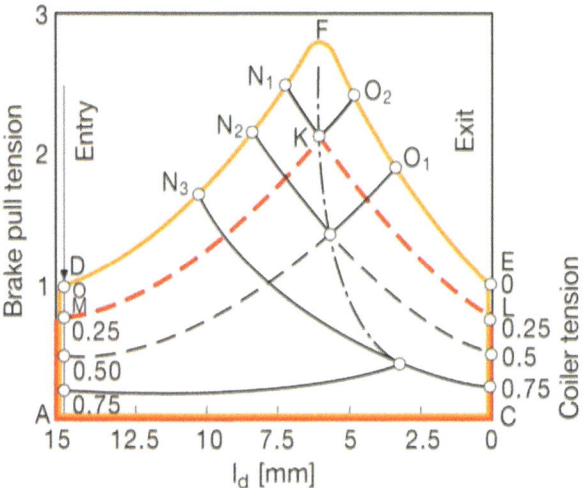

Fig. 10.4: Spezific pressure distribution in the roll gap during rolling with forward and backward tensions /9/

Strip tensions decrease the rolling forces needed, **Fig. 10.4**. The curve segment ADFCA characterizes the pressure trend without applied tensions. When coiler tension, e.g. $0.25 \cdot k_{f_1}$, acts (Section EL), the pressure trend is according to the curve segment ADN_1LCA. If rolling is performed with brake pull tension of the magnitude $0.25 \cdot k_{f_0}$ pressure trend is described by the curve segment AMO_2ECA. In standard cases coiler and brake pull tensions are acting during cold rolling. The pressure trend is then given by the segment AMKLCA if the acting forward tension is $0.25 \cdot k_{f_1}$ and the backward tension is $0.25 \cdot k_{f_0}$. The dashed line gives the pressure maximum or neutral point if both tensions are increased in the same way. Exceeds one of the applied tensions the neutral point is shifted to the opposite direction.

Rolls are released by forward and backward tension, i.e. roll wear is decreased and roll flattening influencing the pressure distribution itself is reduced, too. Due to roll flattening the roll bite length is enlarged on one hand side and the neutral point is shifted to the roll gap exit on the other hand side. These results had been confirmed by rolling experiments.

11. Rolling force, torque and power – general considerations

The rolling (vertical) force per width unit follows - summarizing the vertical stresses along the arc of contact α_0:

$$\frac{F}{b} = r \cdot \int_0^{\alpha_0} \sigma_y \cdot d\alpha \qquad (11.1)$$

or

$$\frac{F}{b} = \int_0^{l_d} \sigma_y \cdot dx . \qquad (11.2)$$

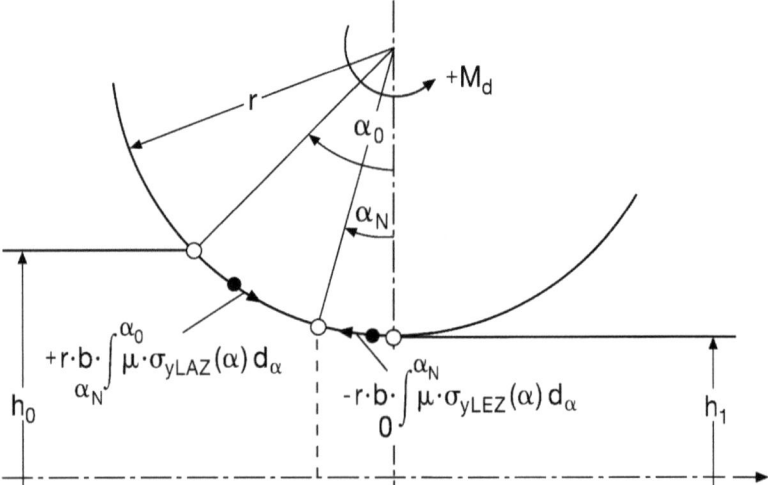

Fig. 11.1: Determining the roll torque for rolling /1, 9/

The roll torque per roll related per width unit is given with r as the lever arm and $\tau_R = \mu \cdot \sigma_N \approx \mu \cdot \sigma_y$ as friction shear stress, **Fig. 11.1**, according to:

$$\frac{M}{b} = r \cdot \left[\int_{\alpha_F}^{\alpha_0} \mu \cdot \sigma_y \cdot r \cdot d\alpha - \int_0^{\alpha_F} \mu \cdot \sigma_y \cdot r \cdot d\alpha \right]$$

$$\frac{M}{b} = r^2 \cdot \left[\int_{\alpha_F}^{\alpha_0} \mu \cdot \sigma_y \cdot d\alpha - \int_0^{\alpha_F} \mu \cdot \sigma_y \cdot d\alpha \right]$$

(11.3)

This is the difference of the resulting friction torques (friction force multiplied with the lever arm) in the lag and forward zone. The friction torques are more or less of the same amount, therefore the accuracy of roll torque calculation depends on the accuracy of the neutral point determination.

The result of the roll force equations using this method can be expressed in general as follows. While for hot rolling

$$\frac{F}{l_d \cdot b} = k_f \left(St, \varphi_h, \dot{\varphi}_h, T \right) \cdot f(Wg, \mu) \tag{11.4}$$

can be used, for cold rolling usually is valid:

$$\frac{F}{l_d \cdot b} = k_f \left(St, \varphi_h, \dot{\varphi}_h \right) \cdot f(Wg, \mu, Z) \tag{11.5}$$

The notation St means the material grade characterized by its chemical composition and its initial state. Wg notes the roll gap geometry and Z means existing strip tensions at the entry and exit side of the roll bite.

Defining k_w as average value of the vertical stress distribution and noting it as deformation resistance it can be written:

$$k_w = \frac{1}{l_d} \cdot \int_0^{l_d} \sigma_y \cdot dx. \qquad (11.6)$$

Therefore (11.2) is equivalent to:

$$F = k_w \cdot l_d \cdot b. \qquad (11.7)$$

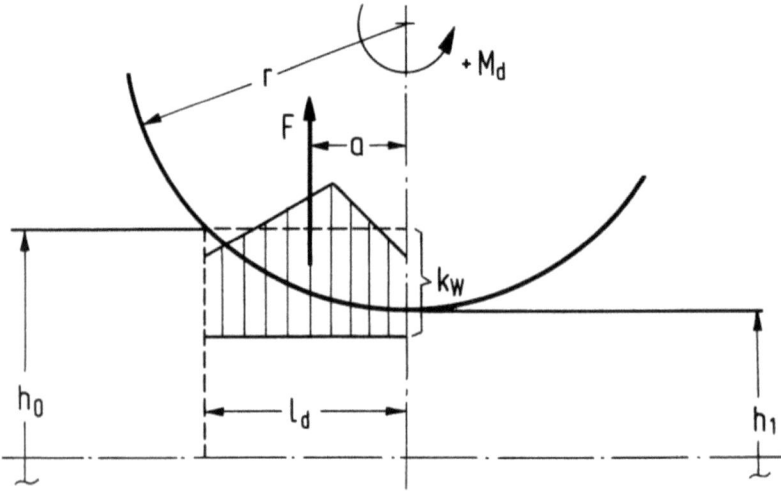

Fig. 11.2: Lever arm method for roll torque calculation /1, 9/

A simplified method for determining the rolling torque is the lever arm method according to Trinks. For one roll it is valid:

$$M = F \cdot a \qquad (11.8)$$

with the lever arm a, **Fig. 11.2**. The lever arm is not known in advance. In practice the lever arm ratio $m = a/l_d$ or $m = \alpha/\alpha_0$ is used. Therefore the total roll torque for both rolls is

$$M = 2 \cdot F \cdot l_d \cdot m. \tag{11.9}$$

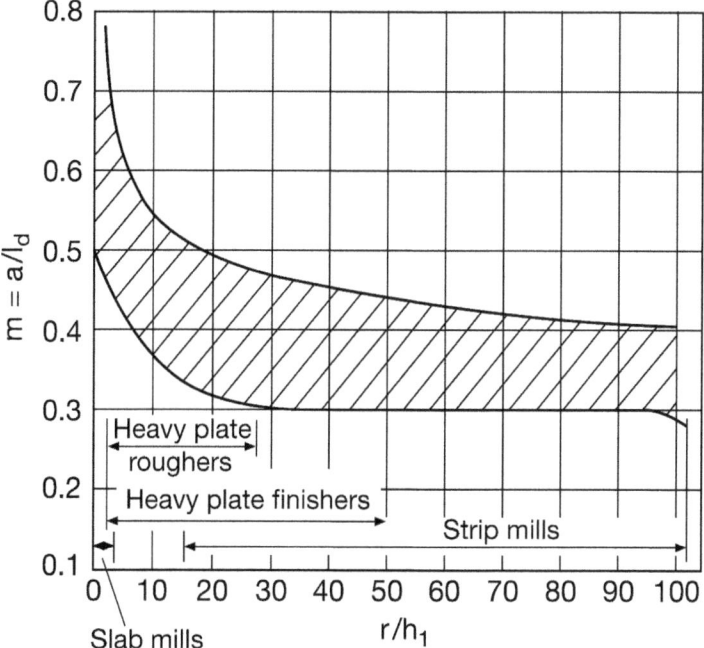

Fig. 11.3: Lever arm ratio for hot rolling of steel without applied tension /1/

For the height ratio $r/h_1 > 10$ the lever arm ratio $m = a/l_d$ for hot rolling of unalloyed steel grades is in between 0.25 and 0.55 and approaches for smaller height ratios asymptotically a value of 0.78, **Fig. 11.3**. For cold rolling of strip the lever arm ratio is in between 0.25 and 0.6, **Fig. 11.4**. For rough estimations in hot and cold rolling m is set to 0.5. The total torque for two rolls can be written according to (11.9):

$$M = F \cdot l_d. \qquad (11.10)$$

Regarding the roll torque M for two rolls and the angular speed ω the roll power follows:

$$P = M \cdot \omega = M \cdot \frac{\pi \cdot n}{30} \qquad (11.11)$$

with the roll revolution n [min^{-1}].

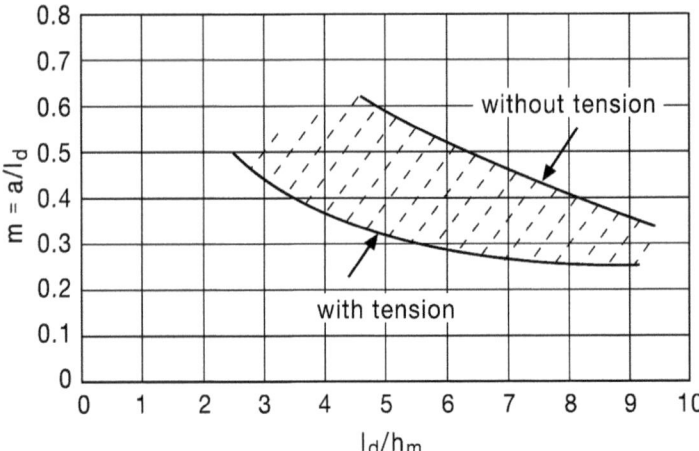

Fig. 11.4: Lever arm ratio for cold rolling of steel with and without applied tension /1/

12. Calculation of rolling force and rolling torque according to Lippmann and Mahrenholtz /5/

An approximation solution of the elementary rolling equation and a reliable calculation method of roll force and roll torque was developed by Lippmann and Mahrenholtz. They made the approach of a power series for the horizontal stress distribution in the roll gap according to the equation

$$\sigma_x(\alpha) = \sigma_x\left(\hat{\alpha}\right) + \sum_{i=1}^{r} a_i \cdot \left(\alpha - \hat{\alpha}\right)^i. \qquad (12.1)$$

In this equation $\alpha = \hat{\alpha}$ denotes a fixed angle coordinate or a fixed cross section in the roll gap respectively. Inserting (12.1) into the elementary rolling equation (10.10) and developing the other occurent functions in power series at $\alpha = \hat{\alpha}$ a recursion formula is provided for the unknown factors a_i. Together with these factors (12.1) provides in general the correct solution for $r \rightarrow \infty$; for finite r due to given convergence in the solution range $0 \leq \alpha \leq \alpha_0$ an approximation is available. Lippmann and Mahrenholtz made a quadratic approach for developing the stress distribution σ_x. Together with the yield criterion $\sigma_y - \sigma_x = k_f$ and the relation between vertical and normal stress

$$\sigma_N = \frac{\sigma_y}{1-\mu \cdot \tan \alpha} \quad \text{for } 0 \leq \alpha \leq \alpha_F, \text{ lead zone}, \qquad (12.2)$$

and

$$\sigma_N = \frac{\sigma_y}{1+\mu \cdot \tan\alpha} \quad \text{for } \alpha_F \leq \alpha \leq \alpha_0, \text{ lag zone} \qquad (12.3)$$

as well as the relation

$$\mu \cdot \sigma_N = k_f/2. \qquad (12.4)$$

For pure sticking friction in the deformation zone they postulate the following equations for the calculation of the vertical stress distribution σ_y for hot rolling:

$$\frac{\sigma_y}{k_f} = \frac{\sigma_F}{k_f} + 1 + \ln\left(\frac{h(\alpha)}{h_1}\right) + \sqrt{\frac{r}{h_1}} \cdot \arctan\left(\sqrt{\frac{r}{h_1}} \cdot \alpha_0\right), \qquad (12.5)$$

$0 \leq \alpha \leq \alpha_N$, lead zone,

$$\frac{\sigma_y}{k_f} = \frac{\sigma_B}{k_f} + 1 + \ln\left(\frac{h(\alpha)}{h_1}\right) + \sqrt{\frac{r}{h_1}} \cdot \left[\arctan\left(\sqrt{\frac{r}{h_1}} \cdot \alpha_0\right) - \arctan\left(\sqrt{\frac{r}{h_1}} \cdot \alpha\right)\right], \qquad (12.6)$$

$\alpha_N \leq \alpha \leq \alpha_0$, lag zone.

In (12.5) and (12.6) σ_B and σ_F denote the magnitude of the horizontal stress at the roll bite entry ($\alpha = \alpha_0$) and the roll bite exit ($\alpha = 0$) respectively.

The related neutral point is calculated by equalizing (12.5) and (12.6):

$$\beta_N = \frac{\alpha_N}{\alpha_0},$$

$$\beta_N = \sqrt{\frac{1-\varepsilon_h}{\varepsilon_h}} \cdot \tan\left(\frac{1}{2} \cdot \sqrt{\frac{h_1}{r}} \cdot \left[\frac{\sigma_B - \sigma_F}{k_f} + \ln(1-\varepsilon_h)\right] + \frac{1}{2} \cdot \arctan\sqrt{\frac{\varepsilon_h}{1-\varepsilon_h}}\right). \quad (12.7)$$

Inserting the solutions for the lag and lead zone in (11.1) and integration using the average yield stress k_{fm} provides the roll force necessary due to

$$F = b \cdot k_{fm} \cdot l_d \cdot Q_F. \quad (12.8)$$

with

$$Q_F = \frac{\sigma_R}{k_{fm}} + 2 \cdot \sqrt{\frac{1-\varepsilon_h}{\varepsilon_h}} \cdot \arctan\sqrt{\frac{\varepsilon_h}{1-\varepsilon_h}} - 1 + \sqrt{\frac{r}{h_1}} \cdot \sqrt{\frac{1-\varepsilon_h}{\varepsilon_h}} \cdot \ln\left[\frac{\sqrt{1-\varepsilon_h}}{1-\varepsilon_h \cdot (1-\beta_N^2)}\right]. \quad (12.9)$$

The roll torque for both rolls can be calculated according to (11.3):

$$M = 2 \cdot b \cdot r \cdot k_{fm} \cdot (h_0 - h_1) \cdot Q_M \quad (12.10)$$

with the function Q_M:

$$Q_M = \sqrt{\frac{r}{h_1}} \cdot \sqrt{\frac{1-\varepsilon_h}{\varepsilon_h}} \cdot \left(\frac{1}{2} - \beta_N\right). \quad (12.11)$$

13. Calculation of rolling force and torque for hot rolling /17/

The process models in many hot strip mills are based on the theory of R. B. Sims developed in 1954. The Sims model including its derivation is given in /17/. Knowing the yield strength of the strip material the effort for calculation is low. Under the assumption of adhering strip material in the roll gap the friction coefficient - the most unknown parameter in the calculations - had been skipped from the calculations. In the following, the equations due to the Sims theory - starting from the elementary rolling equation (10.27) according to von Kárman - are derived. For practical solution of (10.27) further simplifications and approaches are made:

- The friction shear stress, i.e. the product between friction value and normal stress $\mu \cdot \sigma_N$, can exceed the yield strength $k_f/2$ of the material at the contact area between rolls and rolled material; therefore its maximum is given by the yield strength. This assumption is expressed by:

$$\mu \cdot \sigma_N \geq \frac{k_f}{2} \text{ (for } \mu \geq 0.4 \text{)}. \qquad (13.1)$$

If the rolled material sticks to the rolls the shear stress achieves a maximum depending on the yield strength of the material.

- The rolling process is compared with an edging process. With this assumption E. Orowan derived the yield criterion for sticking friction:

$$\sigma_x = \sigma_N - \frac{\pi}{4} \cdot k_f .\tag{13.2}$$

- Roll flattening r' according to Hitchcock's formula (7.1) is considered.
- Material spreading is neglected.
- The horizontal tension stress at the roll bite entry and exit is σ_R and σ_F, respectively.

Under these assumptions the von Kármán differential equation turns to:

$$\frac{d}{d\alpha}\left(\frac{\sigma_x}{k_f}\right) = \frac{r \cdot \pi \cdot \alpha}{2 \cdot (h_1 + r \cdot \alpha^2)} \pm \frac{r}{h_1 + r \cdot \alpha^2} .\tag{13.3}$$

The solution of this equation for the lead zone $(0 \leq \alpha \leq \alpha_N)$ can be written under consideration of the boundary condition

$$\sigma_{xLEZ} = \frac{\pi}{4} \cdot k_{fm} \cdot \ln(h_1 + r \cdot \alpha^2) - k_{fm} \cdot \sqrt{\frac{r}{h_1}} \cdot \arctan\left(\sqrt{\frac{r}{h_1}} \cdot \alpha\right)$$

$$+\sigma_V - \frac{\pi}{4} \cdot k_{fm} \cdot \ln h_1 \tag{13.4}$$

and for the lag zone $(\alpha_N \leq \alpha \leq \alpha_0)$

$$\sigma_{xLAZ} = \frac{\pi}{4} \cdot k_{fm} \cdot \ln(h_1 + r \cdot \alpha^2)$$
$$-k_{fm} \cdot \sqrt{\frac{r}{h_1}} \cdot \arctan\left(\sqrt{\frac{r}{h_1}} \cdot \alpha\right) \qquad (13.5)$$
$$+k_{fm} \cdot \sqrt{\frac{r}{h_1}} \cdot \arctan\left(\sqrt{\frac{h_0 - h_1}{h_1}}\right) + \sigma_R - \frac{\pi}{4} \cdot k_{fm} \cdot \ln h_0$$

When integrating, the assumption of average yield strength k_{fm} of the rolled material according to (9.7) is made.

The neutral angle α_N - and therefore the roll gap position of the stress maximum - can be calculated using the equilibrium condition $\sigma_{xLAZ} = \sigma_{xLEZ}$ resulting in:

$$\alpha_N = \frac{1}{\sqrt{\frac{r}{h_1}}} \cdot \tan\left[\frac{1}{2} \cdot \arctan\left(\sqrt{\frac{h_0 - h_1}{h_1}}\right) + \frac{\pi}{8 \cdot \sqrt{\frac{r}{h_1}}} \cdot [\ln h_1 - \ln h_0] - \frac{1}{2 \cdot \sqrt{\frac{r}{h_1}}} \cdot \frac{\sigma_B + \sigma_F}{k_{fm}}\right]. \qquad (13.6)$$

For tension-free rolling $(\sigma_B = \sigma_F = 0)$ this equation can be simplified to:

$$\alpha_N = \frac{1}{\sqrt{\frac{r}{h_1}}} \cdot \tan\left[\frac{1}{2} \cdot \arctan\left(\sqrt{\frac{h_0 - h_1}{h_1}}\right) + \frac{\pi}{8 \cdot \sqrt{\frac{r}{h_1}}} \cdot [\ln h_1 - \ln h_0]\right]. \qquad (13.7)$$

According to (11.1) the rolling force for hot rolling with only little strip tension (for small rolling angles α the assumption $\sigma_N \approx \sigma_y$ is valid) is given by integration:

$$F = k_{fm} \cdot l_d \cdot b \cdot Q_P\left(\frac{r}{h_1}, \varepsilon_h\right)$$
$$-l_d \cdot b \cdot \left[\sigma_B - \left(\frac{\alpha_N}{\alpha_0}\right) \cdot \sigma_B + \left(\frac{\alpha_N}{\alpha_0}\right) \cdot \sigma_F\right]$$
(13.8)

The function Q_p, **Fig. 13.1**, describes the influence of the roll gap geometry on the roll force under sticking friction conditions.

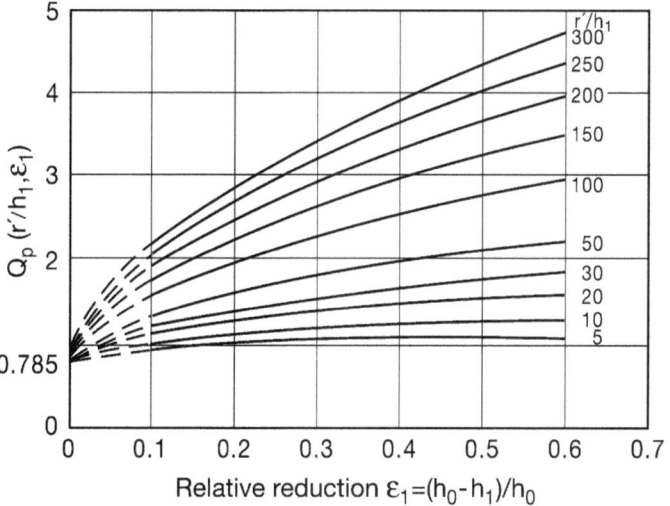

Fig. 13.1: Roll gap geometry function Q_P for roll force calculation in hot rolling case according to Sims /9/

$$Q_P\left(\frac{r}{h_1}, \varepsilon_h\right) = \frac{\pi}{2} \cdot \sqrt{\frac{1-\varepsilon_h}{\varepsilon_h}} \cdot \arctan\sqrt{\frac{\varepsilon_h}{1-\varepsilon_h}} - \frac{\pi}{4}$$
$$- \sqrt{\frac{1-\varepsilon_h}{\varepsilon_h}} \cdot \sqrt{\frac{r}{h_1}} \cdot \ln\left(\frac{h_F}{h_1}\right) + \frac{1}{2} \cdot \sqrt{\frac{1-\varepsilon_h}{\varepsilon_h}} \cdot \sqrt{\frac{r}{h_1}} \cdot \ln\left(\frac{1}{1-\varepsilon_h}\right)$$
(13.9)

For the ratio neutral angle to rolling angle $\alpha_N/\alpha_0 = 0.5$ the roll force is:

$$F = k_{fm} \cdot l_d \cdot b \cdot Q_P\left(\frac{r}{h_1}, \varepsilon_h\right) - l_d \cdot b \cdot \frac{\sigma_B + \sigma_F}{2}. \qquad (13.10)$$

When rolling without strip tension $(\sigma_B = \sigma_F = 0)$ equation

$$\begin{aligned}F &= k_{fm} \cdot l_d \cdot b \cdot Q_P\left(\frac{r}{h_1}, \varepsilon_h\right) \\ &= k_{fm} \cdot \sqrt{r \cdot (h_0 - h_1)} \cdot b \cdot Q_P\left(\frac{r}{h_1}, \varepsilon_h\right)\end{aligned} \qquad (13.11)$$

can be used. This gives a first estimation of the roll forces to be expected.

For calculation of the roll torque (11.3) is considered. The result for two rolls is:

$$M = 2 \cdot r \cdot r' \cdot k_{fm} \cdot b \cdot Q_G\left(\frac{r'}{h_1}, \varepsilon_h\right). \qquad (13.12)$$

The accuracy of the roll force formula is more or less satisfying. Due to the approaches in the Sims model this is not true for roll torque calculation. Therefore for practical use roll torque formula (11.9) is requested which has to be modified for rolling with back and forward tension:

$$M = 2 \cdot F \cdot l_d' \cdot b \cdot m + \left(\frac{\sigma_B \cdot h_0}{2} - \frac{\sigma_F \cdot h_1}{2} \right) \cdot b \qquad (13.13)$$

or with the back- and forward tension forces F_{TB} and F_{TF}

$$F_{TB} = \sigma_B \cdot b \cdot h_0 \qquad (13.14)$$

and

$$F_{TF} = \sigma_F \cdot b \cdot h_1 \qquad (13.15)$$

(13.13) can be written:

$$M = 2 \cdot F \cdot l_d' \cdot b \cdot m + \left(\frac{F_{TB}}{2} - \frac{F_{TF}}{2} \right) \cdot b . \qquad (13.16)$$

For rough approaches in practice it is $m = 0.5$, therewith:

$$M = F \cdot l_d' \cdot b + \left(\frac{F_{TB}}{2} - \frac{F_{TF}}{2} \right) \cdot b . \qquad (13.17)$$

14. Calculation of rolling force and torque for cold rolling

Strip tensions are of major importance in cold rolling mills than in comparison to hot rolling mills because high tensions are applied to effectively reduce the rolling forces and therewith allow higher reductions in the mill. Backward tension is more effective than forward tension. For the roll torque the situation is opposite: Backward tension has to be compensated by the rolling mill stand drive motor and therefore the motor drive capability of the mill stand has to be enlarged. The forward strip tension, e.g. provided by the coiling unit, acts in the same sense as the rolling torque. In this case the rolling mill stand drive unit is supported and more unloaded.

To calculate rolling forces one can find a lot of equations under specific assumptions for specific solutions. The most commonly used cold rolling theory was developed by H. Ford, F. Ellis and D. R. Bland in the 1960ies (BFE-Theory); Roll forces and roll torques can be calculated with a high degree of accuracy and time effectivity, too. Another advantage of the BFE-theory is that the influence of rolling geometry and friction conditions are separated in the formula work and therefore are suitable for automation applications.

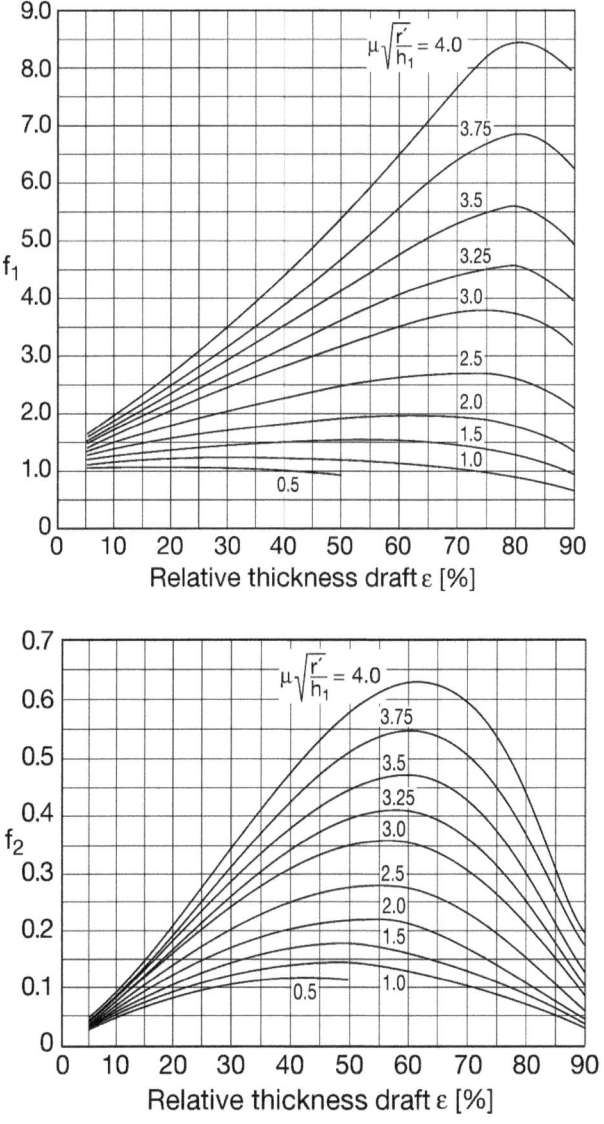

Fig. 14.1: f_1- and f_2-function for determining roll forces and roll torques in cold rolling without applied strip tensions according to Ford and Bland /9/

In the following the equations of the BFE-theory are introduced /9/.

The roll force in cold rolling without applied tensions is:

$$F = k_{fm} \cdot b \cdot \sqrt{r' \cdot (h_0 - h_1)} \cdot f_1\left(\mu \cdot \sqrt{\frac{r'}{h_1}}, \varepsilon_h\right). \qquad (14.1)$$

The roll torque is calculated according to:

$$M = 2 \cdot r \cdot \frac{h_0^2}{h_1} \cdot k_{fm} \cdot b \cdot f_2\left(\mu \cdot \sqrt{\frac{r'}{h_1}}, \varepsilon_h\right). \qquad (14.2)$$

k_{fm} represents the average yield strength between entrance and exit condition, b the strip width, r' the flattened roll radius, h_0 and h_1 the entry and exit thicknesses, respectively, ε_h the relative thickness reduction and r the non-flattened work roll radius.

The functions f_1 and f_2 describe the influence of the roll gap. They are illustrated in **Fig. 14.1**.

The roll force in the cold rolling process with applied strip tensions σ_B and σ_F is /9/

$$F = k_{fm} \cdot b \cdot \sqrt{r' \cdot (h_0 - h_1)} \cdot \left(1 - \frac{\sigma_B}{k_{fm}}\right) \cdot f_3\left(\mu \cdot \sqrt{\frac{r'}{h_1}}, \varepsilon_h, B\right) \qquad (14.3)$$

and the total torque (for both work rolls):

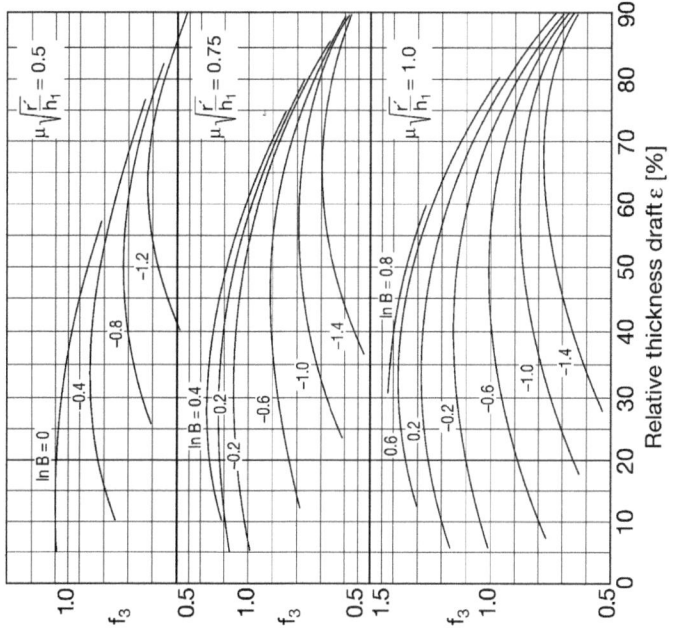

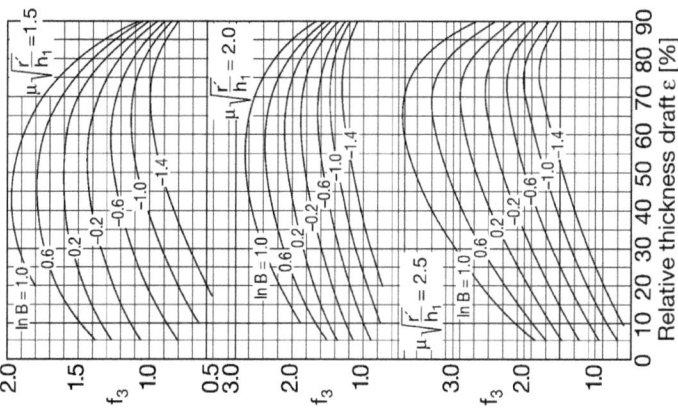

Fig. 14.2: Cold rolling. Function f_3 for the calculation of the roll force and roll torque according to Ford, Ellis, Bland /9/

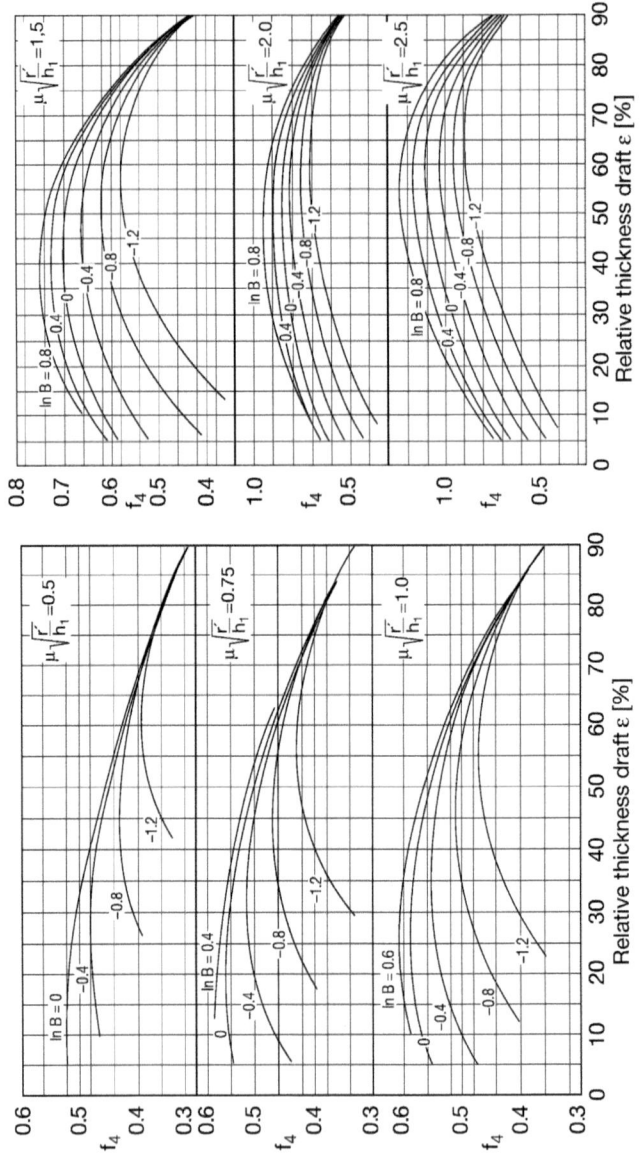

Fig. 14.3: Cold rolling. Function f_4 for determining roll torque according to Ford, Ellis, Bland /9/

$$M = 2 \cdot k_{fm} \cdot b \cdot r \cdot (h_0 - h_1) \cdot \left(1 - \frac{\sigma_B}{k_{fm}}\right) \cdot f_4\left(\mu \cdot \sqrt{\frac{r'}{h_1}}, \varepsilon_h, B\right).$$
$$+ b \cdot r \cdot (\sigma_B \cdot h_0 - \sigma_F \cdot h_1)$$

(14.4)

The strip tension influence is described by

$$B = \left(1 - \frac{\sigma_F}{k_{fm}}\right) / \left(1 - \frac{\sigma_B}{k_{fm}}\right).$$

(14.5)

f_3 and f_4 are dimensionless functions, **Fig. 14.2** and **14.3**. They give the influence of geometry, friction and strip tension.

For practical use the lever arm method according to Trinks is recommended.

15. Examples for the calculation of rolling forces and torques in hot and cold rolling

This paragraph gives examples for roll force and roll torque calculation in hot and cold rolling.

Hot rolling:

As an example a steel strip of width $b = 1000\,\text{mm}$ is taken. Further given data are:

$r = 350\,\text{mm}$,
$h_0 = 10\,\text{mm}$,
$\varepsilon_h = 0.3$,
$k_{fm} = 161\,\text{N}\,\text{mm}^{-2}$ and
$\sigma_B = \sigma_F = 0\,\text{N}\,\text{mm}^{-2}$ (no forward / backward strip tensions at roll bite).

The algorithm of Sims and Trinks is applied for the calculation the rolling force and torque, respectively. First

$h_1 = h_0 \cdot (1-\varepsilon_h) = 10\,\text{mm} \cdot (1-0.3) = 7\,\text{mm}$ and
$l_d = \sqrt{r \cdot (h_0 - h_1)} = \sqrt{350 \cdot (10\,\text{mm} - 7\,\text{mm})} = \sqrt{1050}\,\text{mm} \approx 32\,\text{mm}$

are calculated. For determining the roll force and roll torque the parameters

$\dfrac{r}{h_1} = \dfrac{350\,\text{mm}}{7\,\text{mm}} = 50$ and
$m = 0.37$ (lever arm ratio according to Fig. 11.3)

are essential. According to Fig. 13.1 the roll gap geometry function Q_P is estimated to 1.8.

The roll force and roll torque equations (13.11) and (13.13) provide:

$$F = b \cdot k_{fm} \cdot l_d \cdot Q_P = 1000 \text{ mm} \cdot 161 \frac{N}{mm^2} \cdot 32 \text{ mm} \cdot 2{,}32$$
$$F = 9{,}234 \text{ kN} \approx 9.3 \text{ MN}$$

and

$$M = 2 \cdot F \cdot l_d \cdot b \cdot m = 2 \cdot 9{,}274 \text{ kN} \cdot 0.032 \text{ m} \cdot 0.37 \approx 220 \text{ kNm}.$$

Cold rolling:

As an example for cold rolling an unalloyed steel of grade S235JR is chosen. The given rolling data are:

$r = 250 \text{ mm}$,
$h_0 = 3.0 \text{ mm}$,
$b = 1200 \text{ mm}$,
$k_{fm} = 454 \frac{N}{mm^2}$,
$\mu = 0.1$,
$\Delta h = h_0 - h_1 = 0.9 \text{ mm}$
and
$\sigma_B = \sigma_F = 0 \text{ N/mm}^2$ (no strip tensions at roll gap entry and exit).

The algorithm of Ford, Ellis and Bland without applied strip tensions and without considering roll flattening provides

$$F = k_{fm} \cdot b \cdot \sqrt{r \cdot (h_0 - h_1)} \cdot f_1\left(\mu \cdot \sqrt{\frac{r}{h_1}}, \varepsilon_h\right). \qquad (14.1)$$

With the auxiliary value $f_1 = 1.25$ the roll force is

$$F = 454 \; \frac{N}{mm^2} \cdot 1200 \; mm \cdot \sqrt{250 \; mm \cdot (3 \; mm - 2.1 \; mm)} \cdot 1.25.$$

$F \approx 12.22 \; MN$

The roll torque is given according to

$$M = 2 \cdot r \cdot \frac{h_0^2}{h_1} \cdot k_{fm} \cdot b \cdot f_2\left(\mu \cdot \sqrt{\frac{r}{h_1}}, \varepsilon_h\right) \qquad (14.2)$$

and with the auxiliary value $f_2 = 0.125$ to:

$$M = 2 \cdot 250 \; mm \cdot \frac{(3 \; mm)^2}{2.1 \; mm} \cdot 454 \frac{N}{mm^2} \cdot 1200 \; mm \cdot 0.125.$$

$M \approx 146 \; kNm$

16. Heat balance in hot rolling and temperature calculation

Heat balance in hot rolling generally has to take into account the following aspects:

Temperature losses:

- Radiation losses onto the ambient ΔT_{rad},
- Thermal losses due to heat conductivity to the rolls ΔT_L,
- Temperature losses by controlled water cooling ΔT_{WC},
- Convection losses ΔT_C onto the ambient (neglected in many applications).

Temperature gain:

- Temperature increase due to deformation energy ΔT_{DE}.

The complete heat balance therefore consists of the terms:

$$\Delta T_M = \Delta T_{0M} - \Delta T_{rad} - \Delta T_L - \Delta T_{WC} + \Delta T_{DE} \qquad (16.1)$$

with T_M denoting the actual and T_{0M} the starting temperature of the material.

Analyzing the individual contributions of the heat balance gives:

Temperature loss caused by heat radiation

The radiation heat of a hot body is calculated according to the Stefan-Boltzmann law:

$$Q_{rad} = \varepsilon \cdot C_S \cdot \left[\left(\frac{T_M}{100}\right)^4 - \left(\frac{T_A}{100}\right)^4\right] \cdot t_{rad} \cdot A_S. \qquad (16.2)$$

In the above equation the variables denote the following:

Q_{rad} [J]: Radiation heat,
T_M [K]: Temperature of the rolled material,
T_A [K]: Ambient temperature,
C_S $\left[\dfrac{J}{m^2\, h\, K^4}\right]$: Radiation constant of a black body with $C_S = 5.786\ \dfrac{J}{m^2\, h\, K^4}$,
ε [-]: Emissivity factor for scaled material surfaces ($\varepsilon_{Steel} \approx 0.8 - 0.85$)
t_{rad} [s]: Radiation time and
A_S [mm²]: Radiating material surface.

The radiation heat can be expressed alternatively using

$$Q_{rad} = V_S \cdot \rho \cdot c_P \cdot \Delta T_{rad} \qquad (16.3)$$

with

V_S [mm³]: Radiating volume,
ρ [kg mm^{-3}]: Material density,
c_P [J kg^{-1} K]: Specific heat capacity.

Furthermore the following equations are valid:

$$V_S = A \cdot l \text{ and } U_S \cdot l = A_S \qquad (16.4)$$

denoting

A   [mm²]: Cross section area of the radiating body,
l   [mm]: Length of the radiating body,
U_S [mm]: Circumference of the radiating body.

Equalizing and inserting provides after some conversions and simplifications the equation for temperature due to heat radiation:

$$\Delta T_{rad} = \frac{\varepsilon \cdot C_S \cdot t_{rad} \cdot U_S}{A_S \cdot \rho \cdot c_P} \cdot \left[\left(\frac{T_{rad}}{100}\right)^4 - \left(\frac{T_A}{100}\right)^4\right]. \qquad (16.5)$$

All thermophysical parameters depend on temperature. For basic carbon steel grades rolled in the temperature range between 800 °C and 1,300 °C these data are according to:

Specific heat capacity c_P $\left[J\ kg^{-1}\ K\right]$:

$c_P = c_P(T) = 454.54 + 0.327 \cdot T$

Material density ρ $\left[kg\ mm^{-3}\right]$:

$\rho = \rho(T) = 8.0332 \cdot 10^{-6} - 4.833 \cdot 10^{-10} \cdot T$

Heat conductivity $\left[W\ K^{-1}\ mm^{-1}\right]$:

$\lambda = \lambda(T) = 0.01847 + 8.357 \cdot 10^{-6} \cdot T$

Temperature loss caused by heat conductance to the rolls

The temperature loss due to heat conductance to the rolls had been investigated intensively by Oskar Pawelski /12/. According to the Pawelski model rolled stock and rolls are considered as one-sided infinitely extended bodies separated by a scale layer, **Fig. 16.1**. This model is correct due to the fact that the heat transfer is limited to the both-sided vicinity of the contact area. Therefore core temperatures of the rolls and rolled stock can be assumed constant. The heat conductance temperature loss can be evaluated as following:

$$\Delta T_L = \frac{2 \cdot l_d \cdot b_m \cdot k_L \cdot (T_M - T_{WC})}{c_p \cdot \rho \cdot \dot{V}} \qquad (16.6)$$

with

$b_m = \frac{b_0 + b_1}{2}$ [mm]: Average rolled stock width in the roll gap,

l_d [mm]: Roll bite length,

$\dot{V}$ [mm³ s⁻¹]: Volume flow in the roll gap,

T_M [K]: Temperature of the rolled stock,

T_{WC} [K]: Roll core temperature and

k_L [J K⁻¹ s⁻¹ mm⁻²]: Heat transfer coefficient.

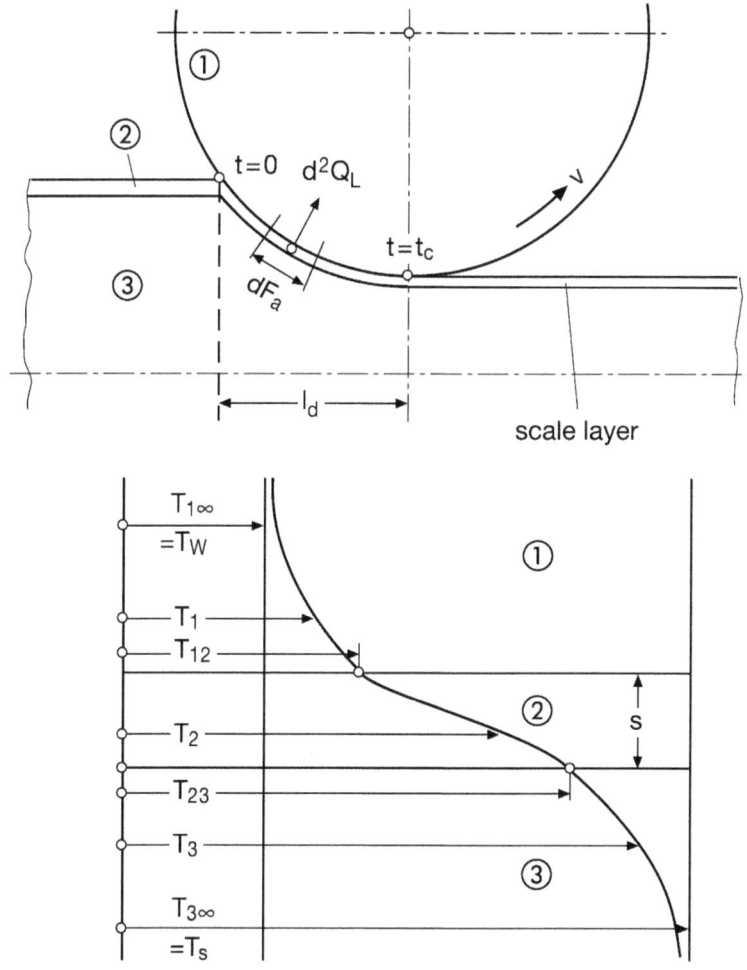

Fig. 16.1: Heat transfer during hot rolling – temperature distribution /12/

The heat transfer coefficient is a complex parameter and depends on scale layer thickness and contact time t_c of the rolled stock in the roll gap, **Fig. 16.2**. For rough estimation two limit values for the heat transfer coefficient can be taken into account:

For the upper limit of the scale thickness ($s_{scale} \to \infty$):

$$k_{Lo} = \frac{8.267 \cdot 10^{-6}}{\sqrt{t_C}} \text{ with } t_C \text{ [s]} \tag{16.7}$$

and for the lower limit value ($s_{scale} \to 0$):

$$k_{Lu} = \frac{2.361 \cdot 10^{-6}}{\sqrt{t_C}} \text{ with } t_C \text{ [s]}. \tag{16.8}$$

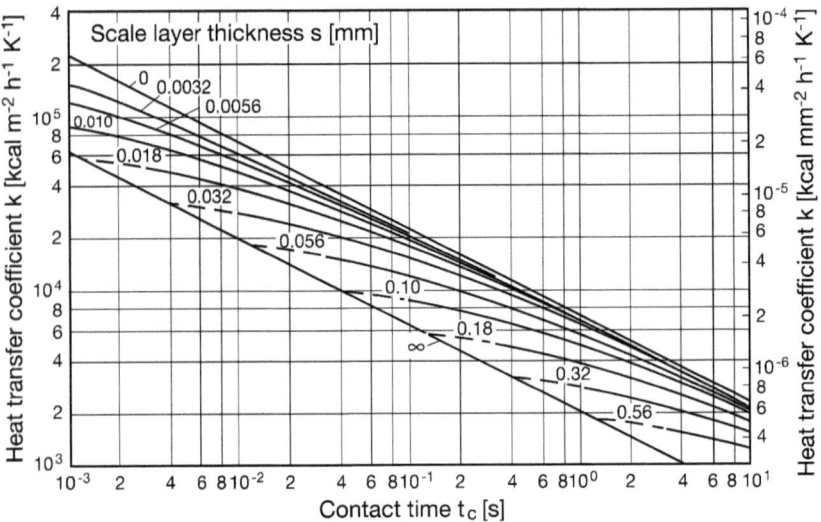

Fig. 16.2: Calculated heat transfer coefficient for hot rolling of a low carbon steel with steel rolls /12/

Temperature loss due to controlled water cooling

The temperature loss due to controlled water cooling can be calculated as:

$$\Delta T_{WC} = T_{Wa} + (T_{SM} - T_{Wa}) \cdot \exp\left(\frac{4 \cdot K \cdot l_{CS}}{c_P \cdot \rho \cdot v_r \cdot b}\right) \quad (16.9)$$

with the variables

T_{Wa}	[°C]:	Water temperature,
T_{SM}	[°C]:	Surface temperature of the rolled stock,
v_r	[m s^{-1}]:	Rolling speed,
L_{CS}	[m]:	Length of the cooling section.
b	[m]:	Rolled width,
K	[W mm^{-2} K^{-1}]:	Heat transfer coefficient.

The heat transfer coefficient K can be calculated according to

$$\frac{1}{K} = \frac{1}{a'} + \frac{1}{a_c} \quad (16.10)$$

with

$$a' = \frac{1}{\sqrt{\pi}} \cdot \frac{\lambda}{\sqrt{a \cdot t_c}}, \quad (16.11)$$

$$a = \frac{\lambda}{\rho \cdot c_P} \; [m^2 \, s^{-1}] \quad (16.12)$$

and t_C as cooling time in s. The parameter a_C denotes the constant heat transfer coefficient with $a_c \approx 10{,}000 \; \frac{W}{m^2 \cdot K}$. For cooling times $t_C > 1$ s the expression $K = a'$ can be used and for cooling times $t_C < 1$ s equation (16.10) is valid for calculation of the heat transfer coefficient K.

Temperature loss due to convection

The temperature loss caused by convection amounts roughly 4 % of the radiation losses.

Temperature gain by dissipation

The temperature increase is calculated using the deformation energy and amounts to:

$$\Delta T_{DE} = \frac{k_{wm} \cdot |\varphi_h|}{c_P \cdot \rho} . \qquad (16.13)$$

k_{wm} denotes the average deformation strength, $|\varphi_h|$ the logarithmic deformation in height direction and c_P and ρ the thermophysical parameters specific heat capacity and material density.

17. Literature

This part gives an overview of the recommended and used literature due to the topic „Basics in Flat Rolling".

/1/ Kopp, R., Wiegels, H.: Einführung in die Umformtechnik. 1. Auflage Aachen, Verlag der Augustinus Buchhandlung, Pontstraße 96, 52062 Aachen, 1998.

/2/ Autorenkollektiv: Rationeller Energieeinsatz bei Umformprozessen. VEB Deutscher Verlag für Grundstoffindustrie, Leipzig 1981.

/3/ Hensel, A., Spittel, T.: Kraft- und Arbeitsbedarf bildsamer Formgebungsverfahren. VEB Deutscher Verlag für Grundstoffindustrie, Leipzig 1978.

/4/ Lange, K.: Umformtechnik.
Band 1 – Grundlagen (1984).
Band 2 – Massivumformung (1988).
Band 3 – Blechbearbeitung (1990).
Band 4 – Sonderverfahren; Werkzeugtechnik, Produktion (1993), Springer-Verlag, Berlin.

/5/ Lippmann, H., Mahrenholtz, O.: Plastomechanik der Umformung metallischer Werkstoffe (1967). Springer Verlag, Berlin.

/6/ Spur, G.: Handbuch Umformen, 2012. Carl Hanser Verlag, München

/7/ Verein Deutscher Eisenhüttenleute: Grundlagen der Bildsamen Formgebung. Verlag Stahleisen 1966.

/8/ Verein Deutscher Eisenhüttenleute: Werkstoffkunde Stahl. Verlag Stahleisen 1984.

/9/ Weber, K. H.: Grundlagen des Bandwalzens. VEB Deutscher Verlag für Grundstoffindustrie, Leipzig 1974.

/10/ H. Hoff, T. Dahl: Grundlagen des Walzverfahrens. Verlag Stahleisen 1955.

/11/ Z. Wusatowski: Grundlagen des Walzens. VEB Deutscher Verlag für Grundstoffindustrie, Leipzig 1963.

/12/ O. Pawelski: Theoretische Grundlagen des Warmwalzens.
Mitteilung aus dem Max-Planck Institut für Eisenforschung, Abhandlung 1174 in: Herstellung von Halbzeug und warmgewalzten Flacherzeugnissen, Verlag Stahleisen, Düsseldorf, 1972

/13/ Stone, M. D.: Rolling of thin strip. Iron and steel engineer 30 (1953) 2, p. 61-74; 33 (1956) 12, p. 55-76.

/14/ Troost, A., Hölling, K.: Berechnung des Kraftbedarfes beim Walzen dünner Bänder und Bestimmung der kleinsten walzbaren Banddicke. Archiv für Eisenhüttenwesen 33 (1962) 7, p. 427-439.

/15/ Stone, M. D.: The rolling of ultra-thin tin plate. Iron and steel engineer 38 (1961) 6, p. 67-69.

/16/ von Kárman, Th.: Beitrag zur Theorie des Walzvorganges. Zeitschrift für Angwandte Mathematik und Mechanik 5, 1925, p. 139-141.

/17/ Sims, R. B.: The Calculation of Roll Force and Torque in Hot Rolling Mills. The Institution of Mechanical Engineers. Proceedings 1954, Volume 168, p. 191-200.

/18/ von Mises, R.: Mechanik der festen Körper im plastisch-deformablen Zustand. Göttinger Nachrichten mathematisch-physikalische Klasse, 1913, p. 582-592.

/19/ Spittel, M., Neubauer, S.: Betrachtungen zur mathematischen Fließkurvenformulierung. Neue Hütte 28 (1983) 1.

18. List of key words

arc element 91
average deformation rate
 .... 48, 50
average flow stress 74, 83
average height 27
backward tension 99
BFE-Theory 115
biting angle 30, 34, 39
biting condition 34, 35
blue brittleness 72
brake pull tension 100
ceramic roll grades 65
coiler tension 99
cold flow curve 72, 76
compression ratio 31
compression stress 90
concave velocity profile 43
condition of continuity .. 46, 47, 52
contact length 64
continuity equation 40
continuous flow curve ... 74
convection 131
convex velocity profile .. 43
Coulomb friction 34
curved crop shear knives
 38
deformation 69, 71, 76
deformation efficiency
 factor 83
deformation energy ... 124, 131
deformation energy
 density 83
deformation energy
 hypothesis 22
deformation heat 84
deformation increment . 24
deformation power 84
deformation rate ... 19, 48, 49, 69, 71, 77, 78
deformation resistance
 103
density 131
deviator stress 22, 24
dislocation density 71
dissipation 9, 131
effective deformation
 energy 83
elastic deformation 15, 16, 24, 68
elastic roll deformation. 63
elasticity module 21
elementary rolling
 equation 86, 88, 106
elementary rolling
 equation according to
 E. Siebel 89
elementary rolling
 equation according to
 Th. von Kárman 94

elementary rolling theory 20, 86
entry speed 52
equilibrium condition ... 19, 86
equilibrium of torque 20
external shear stress.... 94
final rolling speed......... 52
flattened roll radius 117
flow curve..................... 68
flow curves................... 82
flow stress............. 65, 74
forming energy............. 82
forming temperature..... 69
forward tension 99
forward zone 43
friction angle 35
friction coefficient 90
friction free deformation 83
friction shear stress.... 102
friction torque 102
friction value........... 36, 39
frictional force 87
heat conductivity 124
heat radiation 125
heat transfer coefficient 128, 130, 131
height ratio 28
height reduction 16, 28
Hitchcock equation 64
homogeneous compression............. 94
Hooke's law 21, 24
horizontal stress........... 87

horizontal stress distribution 96
horizontal tension stress 110
hot flow curve ... 72, 78, 79
hot rolling theory according to Sims ... 109
hot rolling vertical stress distribution 107
hydrostatic stress 15
ideal deformation energy 83
ideal plastic material..... 71
ideal-plastic body 24
lag zone 43, 87, 88, 89, 92, 107, 110
law of volume constancy 19, 30
lead zone... 87, 88, 89, 92, 106, 110
length ratio 31, 51
lever arm 102, 104
lever arm according to Trinks....................... 103
lever arm method 120
lever arm ratio 104
lever arm ratio cold rolling 105
lever arm ratio hot rolling 104
load shifting.................. 37
local deformation.......... 32
local height 27

135

logarithmic deformation 17, 131
longitudinal stress 98
main normal stress 14
main stress area 14
main stress difference .. 22
material density 54
material grade 102
material hardening 94
material law according to M. Levy 24
material law according to R. v. Mises 24
material softening 71
material strain hardening 69
mill productivity 53, 55
model cold flow curves . 77
multi Roller Stand 66
multi-phase material 72
multi-stage flow curve .. 74
neutral angle 30, 111, 113
neutral point 43, 45, 46, 66, 93, 97, 99, 102, 107
normal component 16
normal stress ... 13, 15, 86, 90, 94, 106
Orowan 110
parallelepipedic deformation 90
phase transformation ... 72
plane deformation .. 40, 86
plastic deformation . 15, 16
Poisson's number 64

precipition 72
pressure distribution roll gap 97
pull through condition .. 38
recrystallization temperature 72
regression analysis 79
related deformation 16
related height reduction 31
related width reduction 31
relative thickness reduction 117
roll bite 96, 102
roll bite area 29
roll bite length 28
roll flattening ... 63, 91, 94, 100, 110
roll flattening constant .. 64
roll force 108, 115, 117
roll force according to Lippmann and Mahrenholtz 106
roll force cold rolling ... 115
roll force equation 102
roll gap 27, 33, 86, 87
roll gap geometry 102, 112
roll gap geometry function 112
roll gap ratio 30
roll power 105
roll revolution 105
roll torque ... 102, 104, 105, 108, 115, 117
roll torque cold rolling 115

roll torque hot rolling .. 113
roll wear 100
rolling angle 30
rolling condition 38
rolling force .. 65, 101, 109
rolling pass 33
rolling time 57
rolling torque 109
scale 129
shear stress 13, 87, 90
shear stress according to Tresca 95
shear stress hypothesis according to Tresca .. 90
side ratio 62
Siebel differential equation 91
sliding friction 66
slip forward 45, 46
slipping zone 43
softening 74
specific heat capacity .. 84, 131
speed up 55
spezific deformation energy 83
spreading 19, 59
spreading formula .. 60, 80
spreading formula according to Bachtinov and Schternow 60
spreading formula according to Geuze .. 60

spreading formula according to Hill 61
spreading formula according to Pawelski 61
spreading formula according to Sander .. 61
spreading formula according to Tafel and Sedlaczek 60
spreading formula according to Wusatowski 61
spreading ratio 31
static friction zone 43
steel factor 81
Stefan-Boltzmann law 125
sticking friction 66, 107, 110
strain hardening 75
strain speed 24
strengthening 71, 78
strengthening exponent 76
stress 12
stress deviator 15
stress distribution .. 86, 89, 106
stress tensor 15
strip bumpings 37
strip slippers 37
strip tension 67, 98, 99, 102
stripe element 90
tensile test 76

total deformation 17
Tresca yield criterion 22
true strain 11
vector 12
velocity 40
vertical stress .. 89, 90, 94, 101, 106
volume constancy . 18, 51, 54
volume element 13
volume flow rate 51
von Kárman 109
von Kárman differential equation 95, 110

von Mises yield criterion 21, 98
water cooling 124, 130
width profile 59
width-to-height ratio 28
wire rod rolling mill 65
yield criterion .. 14, 21, 95, 106, 110
yield factor 54
yield strength 21, 109, 111, 117
yield stress. 21, 66, 68, 69, 90, 94, 95